STOCK CAR RACING ENGINE TECHNOLOGY

ADVANCED ENGINE THEORY AND DESIGN FOR ALL LEVELS OF CIRCLE TRACK RACING

FROM THE EDITORS OF

CIRCLE TRACK MAGAZINE

HPBOOKS

HPBooks
Published by the Penguin Group
Penguin Group (USA) Inc.
375 Hudson Street, New York, New York 10014, USA
Penguin Group (Canada), 90 Eglinton Avenue East, Suite 700, Toronto, Ontario M4P 2Y3, Canada
(a division of Pearson Penguin Canada Inc.)
Penguin Books Ltd., 80 Strand, London WC2R 0RL, England
Penguin Group Ireland, 25 St. Stephen's Green, Dublin 2, Ireland (a division of Penguin Books Ltd.)
Penguin Group (Australia), 250 Camberwell Road, Camberwell, Victoria 3124, Australia
(a division of Pearson Australia Group Pty. Ltd.)
Penguin Books India Pvt. Ltd., 11 Community Centre, Panchsheel Park, New Delhi—110 017, India
Penguin Group (NZ), 67 Apollo Drive, Rosedale, North Shore 0745, Auckland, New Zealand
(a division of Pearson New Zealand Ltd.)
Penguin Books (South Africa) (Pty.) Ltd., 24 Sturdee Avenue, Rosebank, Johannesburg 2196, South Africa

Penguin Books Ltd., Registered Offices: 80 Strand, London WC2R 0RL, England

STOCK CAR RACING ENGINE TECHNOLOGY

Cover design by Tresa Rowe/Bird Studios
Cover photos by *Circle Track* Magazine
Book Editing and Production by Michael Lutfy

First edition: June 2007

ISBN: 978-1-55788-506-7

PRINTED IN THE UNITED STATES OF AMERICA

10 9 8 7 6 5 4 3 2 1

CONTENTS

ACKNOWLEDGMENTS

Special thanks to Jim McFarland, Dennis Wells, Larry Jewett, Chris Jenckes, Carolyn Woodard and Robert Fisher for their contributions. Also thanks to Sean Holzman and Rick Stark of Primedia Enterprises for making this book possible.

INTRODUCTION

Engines, they are the very heart of our race cars. We spend countless hours building, rebuilding, tuning and upgrading them, all in a quest to gain more power that will deliver us to Victory Lane. Today's race engines are technological wonders, complex pieces of machinery that are capable of amazing things. Understanding their systems and how they function are critical to building a winning racing program.

Over the years, race engine technology in the stock car world has taken significant leaps forward. You need to look no further than NASCAR's Nextel Cup Series, where the recent influx of Formula 1 engineers into its ranks is indicative of this advancement. But high-paid engineers invading major-league stock car racing is just one small facet of a growing trend in racing. Each and every aspect of the motor is getting scrutinized. Air filter makers are delivering higher flow numbers through the use of synthetic media. Piston manufacturers are manipulating the skirts to deliver better performance. A cam company has figured out a way to grind a camshaft that delivers a form of traction control. Programmable ignition systems that allow individual cylinder spark timing adjustments are in the reach of the Saturday night racer, and handheld tuners that can act like mini-dynos are just now making their way to the grassroots racer.

As these products and systems continue to evolve, so too must the racer and his/her team. With that in mind, we brought together some of the finest minds in motorsports to compile this book. The subtitle of this book is *Advanced Engine Theory and Design for All Levels of Circle Track Racing,* and that's exactly what it is. Whether you race a Street Stock at Sumter, a Dirt Car at DuQuoin, or a Midget at Madera, this book has the vital engine technology that you need to know and understand in order to build a winning program.

This book is a compilation of the best technical engine articles written for *Circle Track* magazine, the source of advanced technology for motorsports, over the past few years. We arranged them based on subject matter, as if you were building an engine. This easy-to-use format will allow you to quickly refer back to individual chapters in the areas you may be working.

On behalf of the writers and editors of *Circle Track*, we hope that you find it invaluable in building a winning race program.

—Robert Fisher, Editor

Chapter 1

RACE ENGINE BASICS: HORSEPOWER & TORQUE

by Jim McFarland & Chris Jenckes

The NASCAR Nextel Cup engine you see here (left) is not that much different from the engines that run every Saturday night at your local short track (below). The techniques used to make the Nextel Cup engines so durable and consistent are within a short-track, weekend racer's grasp.

The next few chapters will deal with fundamental areas of engine building used in Nextel Cup that can be applied directly to other classes of racing. You will discover new thoughts and ample reinforcements for building consistent, powerful, and competitive "Saturday-night" engines. Jencks lays the hard-core foundation and McFarland provides the theoretical flavor.

There are certain aspects of Nextel Cup engine programs that apply directly to engine building for other classes of racing. To make this clear, we must first examine some of the more important topics. So, first things first ...

GETTING STARTED

Everyone wants to know the secrets of building a race-winning, 800hp Nextel Cup engine. Well, here they are: excellent planning, a thorough understanding of the basics, an unbelievably hard-working and dedicated team, fanatical attention to detail, a testing program second to none, and a budget to support the effort. And even if you can only adopt and apply portions of this set of "secrets," your engine-building program will improve, even if it's just you and your set of tools. No single piece of rocket science will make the power you want. Today's Nextel Cup engine is an ultimate example of the process of continuous improvement.

Before you laugh and begin comparing the multimillion

As indicated in this chapter, evaluating an engine's performance potential can be misdirected simply by comparing peak torque values. It is the "shape" of a given torque curve that can be more descriptive of this potential. More exactly, it is the area bounded by the curve, two rpm points, and a graphical axis (representing an area "envelope") that more clearly predicts engine performance potential—particularly regarding its ability to produce acceleration. As a rule, the greater the area, the greater the performance potential.

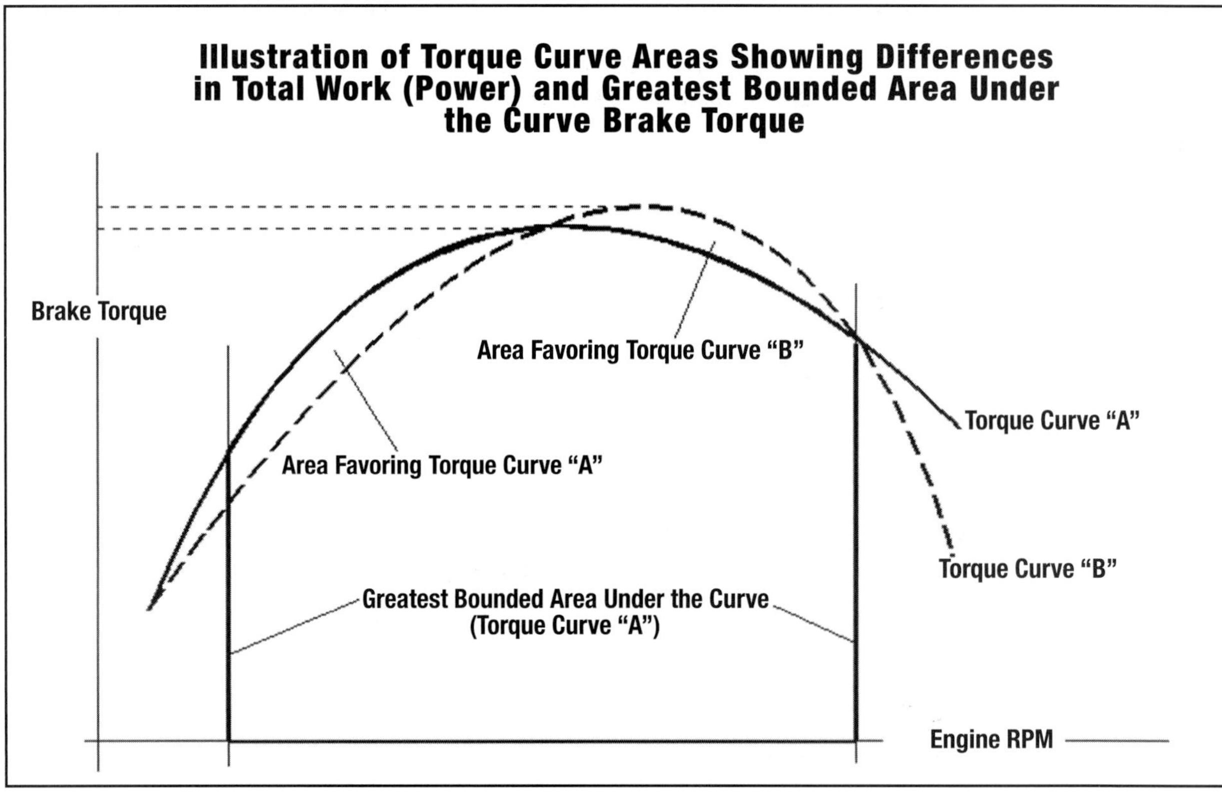

dollar budgets, scores of people, and state-of-the-art equipment to the appearance of your garage or shop, you may be interested in the similarities between successful programs on all levels of racing and how they might apply to your own. For example, a few years ago, a book entitled *The Seven Habits of Highly Successful People* was a bestseller. Its core message was that by adopting common traits of successful people, it's possible to become more efficient and more successful on your own. In a similar fashion, not all of what will be discussed here may be applicable to your particular racing program. But rest assured, the topics will at least be thought-provoking and worthy of consideration.

Many of today's Nextel Cup head engine builders built winning engines at a variety of levels before rising to the top in Cup racing. So what goes on in a Nextel Cup engine shop that could possibly benefit the "Saturday-night" builder?

WHAT IS IMPORTANT?

You must first honestly evaluate what part an engine plays in the entire vehicle performance picture. Further, what part of its performance is the most critical? Circle-track racing is not drag racing. In drag racing, engine output is as important as any single element in the sport. In circle-track racing, engine output alone is probably not even one of the top three factors required for a winning car. Engines are important, but any Nextel Cup head engine builder will say that durability and consistency are the key performance factors for winning races and championships, not engine power.

When it comes to engine performance, it's production of torque and horsepower created in the rpm range that the driver will use the most. This is an important point to research and will be key information in the determination of what and how the engine will be built. For example, if the engine is to be raced at many different tracks and the rpm range varies widely, a much more flexible combination (in terms of rpm range) must be built. Also important, don't forget to take into account how much the rpm range drops on long runs due to tire wear.

TORQUE VS. HORSEPOWER

Technically speaking, torque may be defined as the "potential" to do work. Grab the knob of a locked door, apply a twisting force, and the knob doesn't turn.

This is torque. Upon unlocking, the knob will turn under force of the torque applied, time enters the picture and horsepower can be measured (torque acting over a period of time: pounds-feet/second or pounds-feet/minute). Stated another way, torque is the ability of a system (engine) to perform work. Horsepower is the rate of doing work.

Torque accelerates a race car (the rate of change in torque); horsepower makes it fast (the ability to

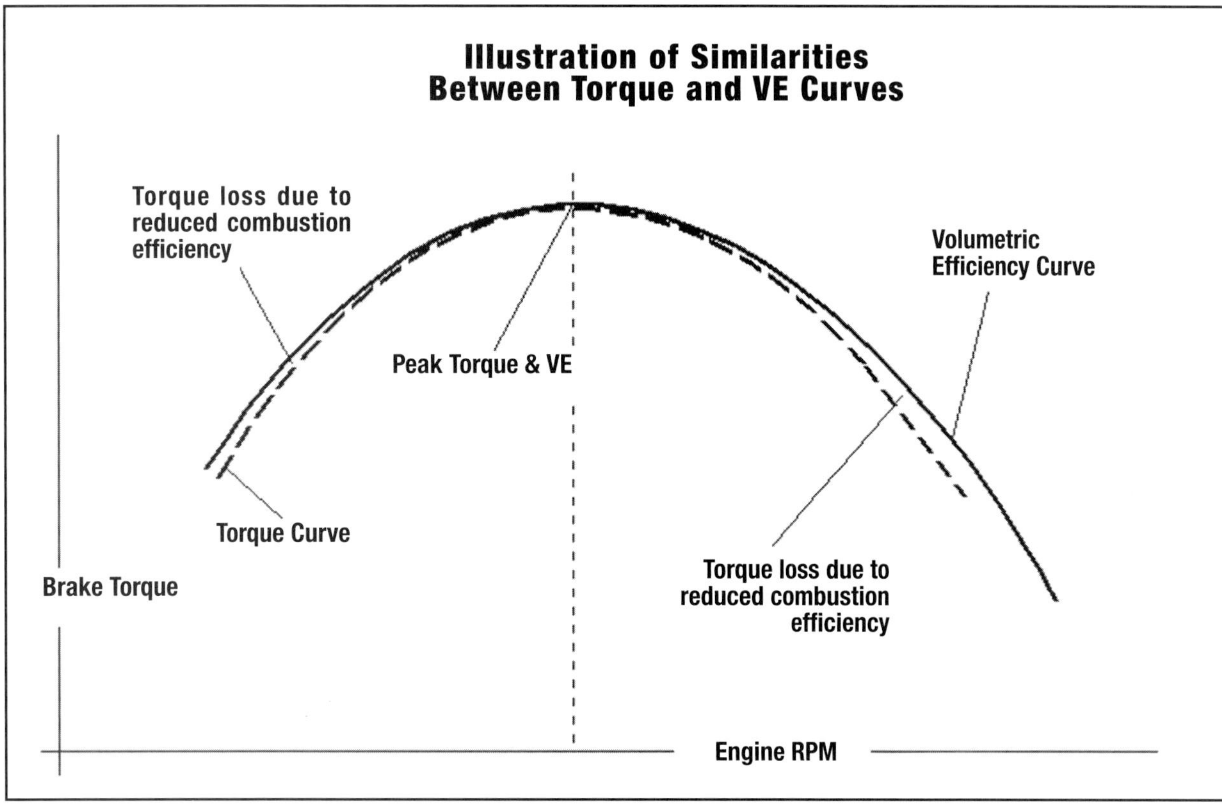

While there is similarity between an engine's "air capacity" and "volumetric efficiency" curves, the relationship of VE and torque curves is an important engine building concept. Below peak torque, at which combustion efficiency normally tends to be optimum (noted frequently by low wide-open-throttle Brake Specific Fuel Consumption data), poor air/fuel mixing reduces combustion efficiency. Above peak torque, air/fuel separation can contribute to lost power, thereby causing differences between VE and torque curves at these rpm.

sustain specific levels of torque). Many engine builders (Nextel Cup and Saturday-night alike) prefer building engines keyed on the production of torque.

An important ingredient in understanding the concept of torque is an engine's ability to produce torque over a specific range of rpm at a rapid rate.

Some call this "transient" torque or the rate at which an engine will accelerate under load through a range of rpm. Generally speaking, the higher the torque output of a given engine, the quicker it will accelerate under load. This is because its "potential" for rpm increase is linked to total torque output.

Peak Torque vs. Area Under the Torque Curve

Based on a previous definition of torque that it is the "potential" for doing work, it is the area under the torque curve that will indicate the magnitude of this potential (see illustration above). Here, it's critical to differentiate between a value for peak torque and a value for the area computed (or shown) below a given curve.

At the risk of oversimplification, we are dealing with an engine's ability to produce power, based on its torque potential. In other words, the greater an engine's torque potential over a range of rpm (as represented by the area under the torque curve), the greater its ability to produce race car acceleration.

As opposed to the importance of a value at peak torque, it's possible to produce quicker acceleration by increasing the area under the torque curve (a "broader" torque band), even though there may be an accompanying slight decrease in peak torque (see illustration on pg. 2).

Equally important are the benefits gained from selecting engine components (parts "integration") that are functionally intended to enhance torque in a predetermined range of engine speed. It is from this base of power that other vehicle components (particularly the powertrain) can be suitably matched to track length and conditions.

Start With a Plan

Know the rules. In general, circle-track racing engines (from Nextel Cup to your local track) are tightly regulated to manage both engine output and cost. You must know the rules intimately in order to build the best engines possible within the rules. Analyze rules carefully and put emphasis only in the area(s) where the greatest return is possible. For example, if you are running an engine in an intake-restricted class that creates airflow limitations (such as a 2-barrel carburetor), use just enough valvespring to keep the valvetrain stable since more spring will absorb horsepower with no benefit.

How long must the engine last (how many races or miles)? How much maintenance will be performed or required? What does the power curve need to look like? (There are many shapes to a power curve, all of which may exhibit the same peak power.) As an example, an engine that is raced in a lightweight car on dirt can afford to sacrifice torque at the corner exit speed for power at the flag stand. A heavier car on pavement must have more torque for corner exit speed to help with acceleration.

And don't forget the budget. Even Nextel Cup engine shops have a budget they must work within. Your planning must include how much time and money you can spend. Building successful racing engines is about using your resources of time and money better than your competition.

WHERE IS THE POWER?

Power comes mainly from the valvetrain, cylinder heads, and intake manifold, no matter what type of race engine you are building. So when looking for more power, this is the place to spend the time and available money. The trend in circle track racing is similar to most other forms of motorsports: more engine speed and higher numerical gear ratios. Crankshaft speed and gear will make more torque at the tire's contact patch than crankshaft torque by itself. The multiplication factor that gear gives to engine torque is what provides you the advantage. The effect is similar to putting an extension arm on a breaker bar when breaking something loose. The downside of this trend, however, is it's hard on valvetrain parts and difficult to pursue (time and cost).

Increasing Volumetric Efficiency

Part of an engine's ability to make power depends upon how efficiently fuel and air can be induced on each inlet cycle. The amount of useful fuel it can ingest depends upon how much air (oxygen) can be supplied during this same time. Therefore, the limiting factor to the amount of power (work) that can be produced is the "air capacity" of each cylinder.

Now, depending upon several factors (many of which involve parts or systems that affect intake airflow), the goal is to try to achieve the highest levels of cylinder filling efficiency as possible—throughout the rpm range.

Here's how all this is tied to torque. The greater the level of cylinder filling efficiency—which, for purposes of this discussion, we'll call "volumetric efficiency" (VE)—the greater the potential for optimizing torque. In fact (as noted in the illustration on pg. 3), an engine's VE curve and torque curve very closely resemble each other. Combustion efficiency influence notwithstanding, it's a pretty good rule of thumb that volumetric efficiency increases represent corresponding gains in torque—again a reflection of the similarity between the two when represented graphically.

In studying the illustration of VE and torque curve similarities, note that the torque curve lies beneath the VE curve both below and above peak torque rpm. Generally, this is caused by poor air/fuel mixture quality. Below torque peak, a lack of flow velocity (related to air/fuel mixing) reduces combustion efficiency (power). Above peak torque, either mechanical separation of mixtures, combustion residue (exhaust gas), or a combination of both, also tends to penalize power. Later on in the next few chapters, this issue will be examined more closely—particularly regarding parts selection and/or modification.

More to the point, because of their influence on volumetric efficiency, the selection of cylinder heads, camshafts, and intake manifolds is critical to not only torque output but also the production of torque within specific ranges of engine speed. Because more airflow does not always equate with increased power, both flow volume and quality must be considered in choosing these parts. This is especially true of cylinder heads and intake manifolds. More on this topic will be discussed as the book later digs into the essentials of these components and the similarities between their application in Nextel Cup and Saturday-night engines.

Torque Multiplication

It is by the application of gears and their respective ratios that mechanical advantages can enhance the application of power to the track. It was mentioned earlier that the rate of torque application (torque delivered vs. delivery time) can improve race car acceleration. By the method of tooth number and diameter differences, degrees of "leverage" can be created that allow an increase in the rate at which torque is applied to the driving wheels.

Even if only one or two forward transmission gears are used, proper selection of gear ratios that allow quick acceleration into the desired span of rpm most often used on the track helps keep a race car competitive. In the consideration of gear ratio (torque multiplication) choices, matching track length and surface conditions to the ratio that allows acceleration into the desired rpm span is critical. For example, an engine with a narrow torque band that will be run in a high rpm range will benefit from a ratio that allows it rapid

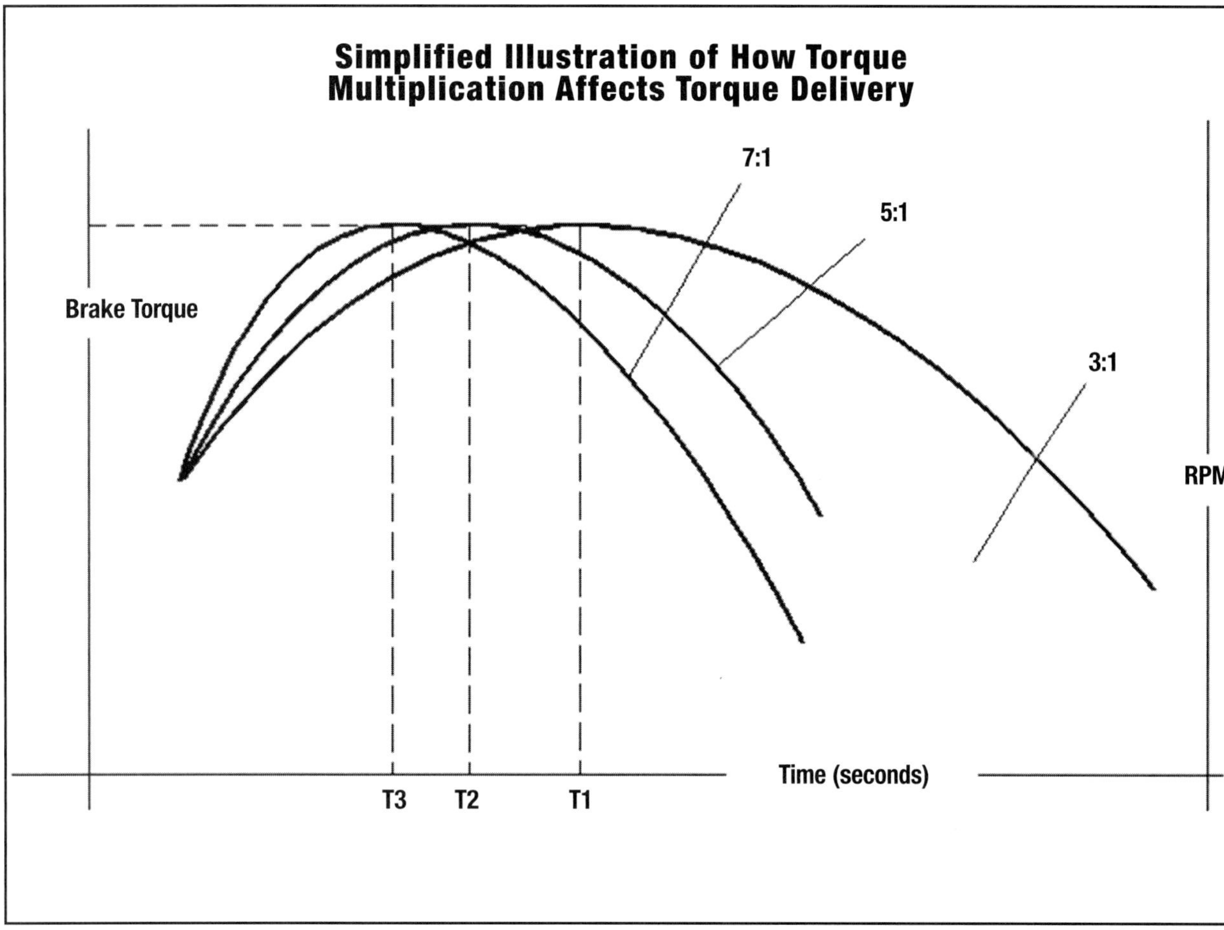

Greatly simplified, this illustration shows how the torque delivery time (as a function of torque multiplication or gear ratio) is shortened as mechanical advantage is gained by a change in ratio. For purposes of illustration and while peak torque rpm and the value attained at this point may remain unchanged, the time required to reach these points is shortened, thereby taking advantage of a given torque.

acceleration up into that range. Larger displacement engines with more area under the torque curve, and at a lower rpm, can pull lower numerical gears and still provide acceleration based on sheer torque.

It is also worth remembering how tire wear affects on-track rpm. Generally, as tires become worn, it becomes necessary to race with less throttle opening (to maintain traction). This reduces the rpm in which the engine runs, thereby causing it to fall lower on its torque curve. Gear selection (torque multiplication) that factors in this condition (particularly on asphalt surfaces) will allow improved torque delivery as traction is reduced by tire wear.

Quality Control is the Key to Durability

Every part, every clearance must be checked every time. Do not ever assume that a part is dimensionally acceptable now, just because it measured correctly when it was purchased or the last time it was checked. Do not let other people's dimensional problems become your engine failure. Do not check piston-to-valve clearance on one cylinder. Check them all. Just as flying has been described as hour after hour of boredom with a few moments of sheer terror, building race engines often involves long hours of incredible repetition (the checking and rechecking of dimensions), brightened with a few hours of excitement when it finally hits the dyno or racetrack.

Keep Records—A notebook containing all of the information on its build should accompany every engine. This should include every important dimension and clearance in the engine with parts lists, dyno time (if applicable), and the amount of run time (even time on the track).

Build Them the Same—In Pro Stock drag racing, it is common that each engine build is slightly different from every other, in a continuing search for more power. Most circle-track engine programs will benefit from consistency, build to build and engine to engine. A tightly controlled recipe yields engines of similar power. When an improvement is (or can be) found, the recipe is updated. For racers who only build one or two engines per season, freshening whenever required, the need for careful parts selection and combinations becomes even more critical to success. Choose parts and make changes carefully, only after serious consideration of all the consequences of such choices.

Stay Focused, Avoid the "Tweak of the Week"

Hard work and good testing will see you through, not the latest fad. Just because someone else is running it doesn't necessarily mean it's right for your combination. The valvetrain is one area where you see this problem, again and again. You hear someone is making great power with a new cam, but it may not work in your engine package. It would be unwise to ignore what is going on around you, but keep it all in context. The best engine programs are the ones that steer their own course, guided by sensible parts decisions and assembly techniques, not simply following the path of someone else.

The top Nextel Cup shops do one thing better than any other form of racing in the world. Yes, even better than all the mighty F-1 programs in Europe and Asia. That is testing. It is routine for a Nextel Cup engine to be able to see variances of 0.5 hp out of 800. That amounts to a resolution of 0.0625 percent and is simply incredible. This allows a head engine builder to separate truth from fiction, as far as changes or improvements are measured.

Chapter 2

RACE ENGINE BASICS: THE BOTTOM END

by Jim McFarland & Chris Jenckes

This chapter on race engine tech deals with the short-block and its components. The NASCAR Nextel Cup engine shown at left benefits from meticulous parts selection and has excellent repeatability from engine to engine. Your short-track engine (below) can benefit from the same meticulous preparation.

THE BLOCK

The cylinder block is the foundation for the engine, so inspection of the casting is critical. Sonic-checking is a service available at most machine shops today and is critical to ensure stability of the block. The small-block Chevy and small-block Ford each have areas where the castings are traditionally thin. You should rely on someone with experience with your type of engine to check the block to make sure it is acceptable. For example, a small-block Chevy is traditionally thin at the base of cylinder No. 7 and should never be any thinner than 0.140 inch for Late Model Stock applications. Engines with less power can afford to be 0.125 inches at the minimum.

Lifter bores must be checked for proper position since "misplacement" or poor "indexing" is common in as-cast conditions, thereby affecting valve timing and valvetrain geometry. Blocks that were not specifically designed for motorsports applications are particularly problematic. Line bore and hone the mains to proper size and make sure the deck is at a true 90-degrees-to-the-crank centerline. If the block you have is of unknown history, it is also a good idea to check the distance from the crank centerline to the cam centerline because this can move (or change) with repeated line boring.

Cylinder deck height should also be checked. The deck height can be adjusted within a small window to optimize a

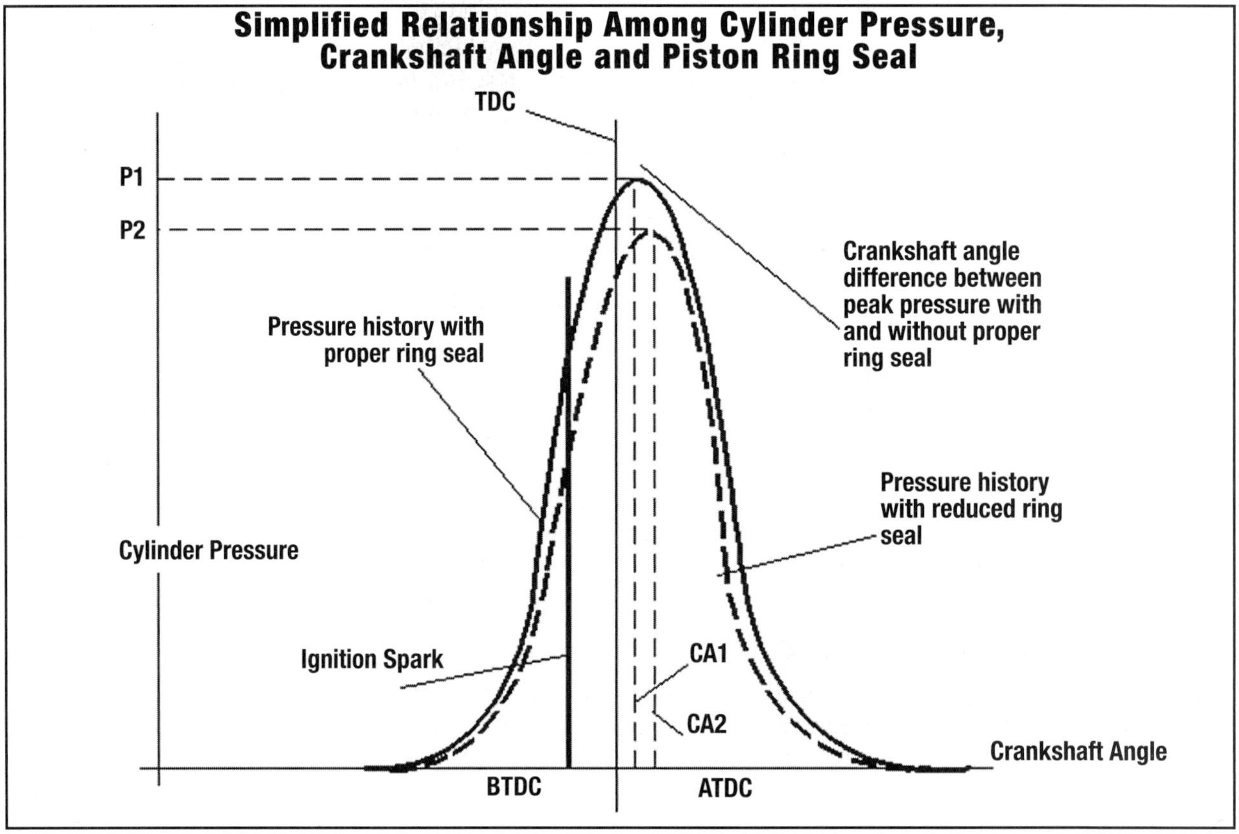

Take a moment to study this illustration. Typically, peak cylinder pressure occurs slightly past TDC piston position, delaying further with increases in rpm. As ring seal deteriorates, both peak cylinder pressure (P1, P2) and the crank angle (CA1, CA2) at which it occurs is also decreased. See story for additional comments.

given engine configuration. For example, when hood height severely restricts the induction system performance, running the minimum deck can create some room. This is usually not an option for budget engine builds. Deburr the block and lighten it as much as the rules allow.

CRANKSHAFT

Buy the best parts that fit into your budget. Higher engine speeds or an engine that must run a full season without a rebuild require better parts. Remember, when working on a limited budget, the money should go into valvetrain, cylinder heads, and intakes for power. Lighter cranks will allow the engine to accelerate faster and can ease the load on the bearings, but they tend to cost more and may have a shorter service life.

How Rotational and Reciprocating Mass Affect Engine Acceleration

Think of rotating crankshaft and reciprocating piston/rod assemblies as flywheels. The greater the total mass, the more resistance offered to changes in system acceleration (or deceleration). Also, the shorter the stroke, the less the mass moment of inertia or resistance to a change in rpm. In the so-called "slider-crank" mechanism of the crankshaft (weight and stroke length), connecting rod (length and weight), and piston (weight and pin position), changing the energy state of this system is a function of weight and leverages. Keep in mind that changes of rotational or reciprocating speed requires energy that might otherwise be available to accelerate a race car.

Journal Sizes

The trend in most forms of motorsports is to run smaller and smaller journal sizes where rules permit. This is not for budget builds but will benefit high-speed engines the most. The reduction in journal radius reduces the speed of the journal at any engine speed and can reduce friction. NASCAR has limited rod journals to a minimum of 1.847 inches. If this size will work to support an 800hp Cup engine, it may be suitable in yours. Only forged cranks specifically designed for these journals can handle the loads, so don't try cutting down your cast crank to run Honda bearings.

Journal Size and Bearing Speed—It's a complex subject. Simply stated, there is a trade-off among features of strength, durability, weight, heat and friction. Small journals reduce friction horsepower losses and present an opportunity to run smaller journals and lighter weight rods. They accelerate more quickly but also offer less strength, higher unit loadings, and increased oil temperature (at the bearing surface interfaces). Typically, the highest unit loading occurs at or just past TDC (depending upon combustion conditions), and the greatest rate of change in surface friction appears at or near maximum piston velocity (in the range of 60–75 degrees past TDC during the power stroke). Generally, engines required to operate over a wide range of rpm seem to benefit more from smaller connecting rod journals than those in a narrow span of engine speed. In making decisions about journal size, the larger the diameter, the greater the bearing speed (relative motion between journal surface and bearing surface during any interval of rpm). As journal diameter is decreased, unit loading (psi) increases with bearing surface area, oil temperature can rise (cooling ability decreases) while the rotational weight reduction aids crankshaft acceleration. Consider these and related effects before making a change.

CONNECTING RODS

Again, buy the best parts that fit into your budget. Higher engine speeds require the best parts. Even budget builds should use premium rod bolts.

Pin Oiling

Some rod manufacturers offer a modification where an oil passage is bored by electrical discharge machining (EDM) from the journal though the beam of the rod to oil the pin. The oil passage in the rod supplies pressurized oil to lubricate the pin. This type of rod should only be considered if pin wear or galling is observed in previous builds or is a concern. If the EDM operation is not performed with care, the process may initiate cracks. Adding pressurized pin oil will increase windage losses.

Wristpins

Wristpin problems are common in high-rpm, high-horsepower engines. The loads placed on the pin tend to cause two primary types of distortion: pure bending and hoop bending. Pure bending, as the name implies, tries to bend the pin along its length. Hoop bending tries to distort the pin radially in the pin bores of the piston.

Pins are an example of where modulus is more important than mass. If the pin is not stiff enough, at best it will deform and cause power to be consumed by friction. In a worst-case scenario, hoop bending will cause the pin to stick in the pin bore, or pure bending will cause the pin to seize in the small end of the rod.

The key is to put the mass in the right place. Increasing the overall diameter will improve the pure bending performance of a pin. Reducing the overall diameter while increasing wall thickness will improve the hoop bending performance of a pin. Obviously, it is not possible to do both, so it is important to understand how the pin is being loaded to pick the best part. Work with your piston and rod supplier to better understand these issues and find the best pin for your application. This can be a critical part.

What is Modulus and Why Is It Important?

Here, this term refers to a modulus of elasticity. By one definition, it is a mathematical ratio of unit stress to unit deformation of a material and is a constant so long as the material is not loaded into permanent deformation (proportional limit). What does this mean to an engine builder? It means that parts (not always metal) under repetitious loading will return to their original shape once the load is removed. Sometimes, this is called "memory." Comparing an example sample of steel to aluminum, the value of E (modulus of elasticity) for steel might be 29,000,000 psi and only 10,300,000 for aluminum.

In engine building (particularly parts selection), there are times when the ability of a given part to be repeatedly loaded within its limit of elasticity is more important than the static (or dynamic) weight of the part. For example, lightweight piston pins that are operated outside their modulus of elasticity have a higher cyclic failure probability than, perhaps, a heavier pin that is continually loaded within its modulus of elasticity.

Pin Coatings

Diamond-like coatings (DLCs) can be very effective in preventing pins from galling the pin bores. When coated with lubricant, these coatings have an extremely low coefficient of friction. Not all DLCs are the same and coating retention can become an issue. It is best to purchase the pins already coated. Coatings are a last line of defense and should be used as such, not relied upon to fix a design problem.

THE CYLINDER KIT

The cylinder kit is the system of parts, including the piston, rings, and cylinder bore, which serve as components working together to seal the combustion gasses from the sump (crankcase) and transfer the chemical energy (combustion) into mechanical energy (torque) with as few losses as possible. The pin and connecting rod are generally included in this system.

In order to minimize friction and maximize

Crank angle "A" represents the period of highest connecting rod loading. Since detonation typically begins just prior to TDC, abnormal bearing pressure can occur nearer TDC than when normal (maximum) cylinder pressure develops over the span indicated. Maximum relative motion of connecting rod journal and bearing surface (bearing "speed") typically occurs during span "B" (see text).

Illustration of Crankshaft Locations Where Connecting Rod Bearing Loadings Are Critical

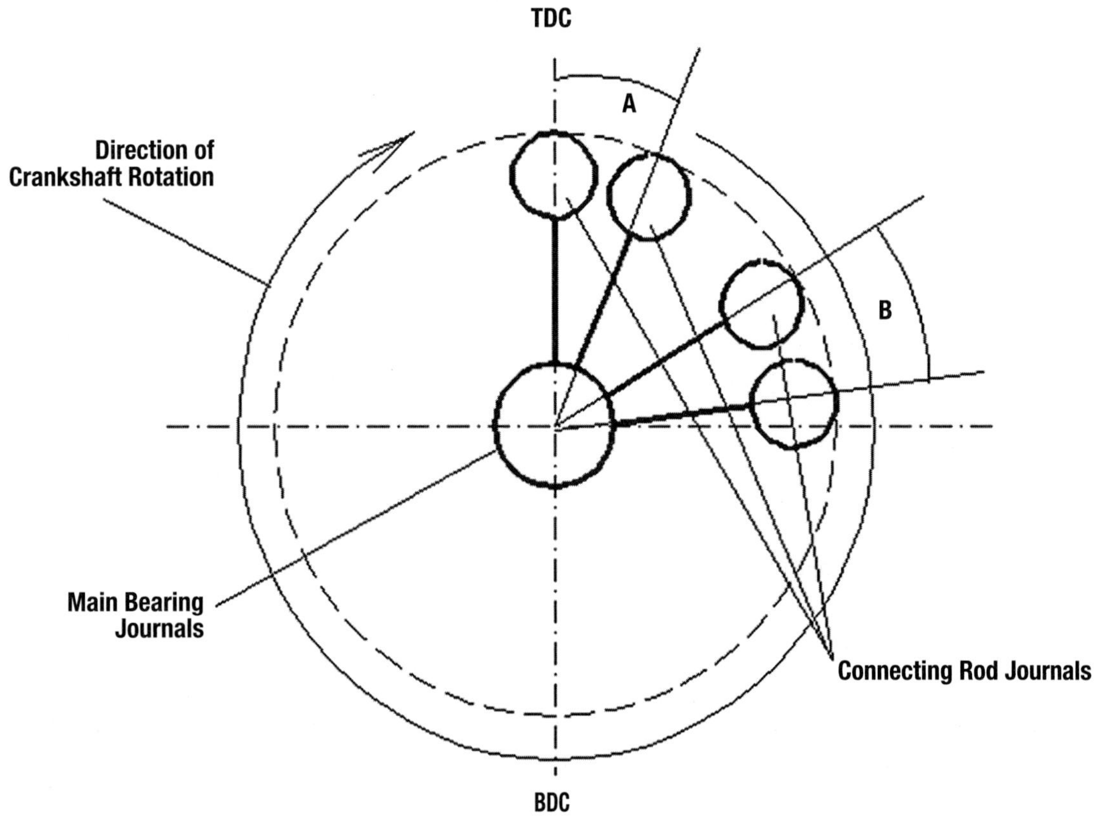

power we must look at the piston, rings, and cylinder bore and their interaction as one system, since a change in one component can influence the entire system. Friction reduction in the cylinder kit area can lead to significant power gains. The ring/piston/cylinder wall interface is the single largest source of friction in the engine. Even small reductions in friction in this area can lead to measurable gains in power.

Cylinder Bore

The bottom line here is that you can't get the bore round and straight enough. Improving concentricity (roundness) of the bore itself will not reduce friction or improve ring seal but it will enable the rings to seal more efficiently and allow a piston skirt design for skirt-to-wall friction reduction.

Benefits of stress plate honing are well documented. This is the first step in improving concentricity. Hot honing is an extension of this process that attempts to duplicate the running state of the block as it is honed with coolant passing through the deck plates and water passages. The cutting fluid is also heated. In Nextel Cup applications, bore distortions of 0.0015 inch have been documented when increasing blocks from room temperature to 220 degrees F.

With the hot-honing process, bore distortion (at temperature) can be reduced to 0.0003–0.0005 inch, a significant reduction. When the cylinder bores are hot honed, they are round at operating temperature but return to an out-of-round condition at room temperature. Sufficient piston-to-wall clearance is maintained at room temperature to allow for engine start-up clearances.

Hot honing itself may reduce friction on production-type cylinder blocks where the temperature distortions are outrageous. Aluminum blocks can benefit significantly from this process. Most well-prepared racing engines may not be able to measure the benefit from hot honing itself. However, when combined with an optimized piston, this process can provide measurable gains.

Nikasil is an extremely hard coating that has been used on cylinder bores in two-cycle engines for years. It can be applied directly to the cylinder block or applied to sleeves, which are then installed and honed to finish. Both parent metal coating (applied directly to the bore) and sleeves have their problems, so this technology is not for Saturday nighters on a budget or the faint of heart. Engine builders have been experimenting with Nikasil for years. While claims have been made for friction reduction, it is my (Jenckes) opinion that Nikasil's primary benefit is the hardness of the coating it produces.

The hard face of a Nikasil bore will allow different types of ring face coatings and ring materials to be used, all of which can lead to measurable friction reduction. Nikasil is tricky to hone, and a power loss can result from reduced ring seal if the honing process is not correct. Hot honing is a good process to use with this type of coating. With the appropriate ring material and ring face coating, it is likely that the bore and ring package can be reused from build to build just by replacing the pistons.

PISTONS

In recent years, piston technology has undergone a revolution with almost all major manufacturers bringing new designs intended to reduce mass, friction, and improve ring seal. Much of this technology comes from Formula 1 and Indycars. These lightweight piston designs include shorter skirt slippers and can have significantly less inertia and friction than more traditional designs.

The piston is also a key in developing ring seal. Ultra-flat top ring lands can reduce "micro welding" or the transfer of material from the piston to the ring. Micro welding will usually cause a complete loss of ring seal and scuff the bore. Any machining operations that are performed to the piston after the top ring groove has been cut can affect the flatness of the land. This should be avoided if possible.

Nextel Cup teams and local racers can both benefit from pistons designed specifically for their particular application. The more a piston design departs from the traditional full-skirt design, the more important all of the details in the cylinder kit become. For example, with less skirt comes higher area loads. As the skirt length is reduced, piston contact areas move upward on the piston. As material is removed for mass reduction, pistons can become susceptible to over temping (excessive heat) from lean air/fuel mixture operation, possibly making it brittle and prone to failure.

The rule of thumb for pistons is to work with a good supplier experienced with your type of racing to determine if these new designs are correct for your particular application. New technology is normally good. Just make certain it fits into your program. Never be reluctant to ask questions.

Gas Porting

Gas porting can improve both ring seal and durability by allowing the combustion pressure to push the top ring against the cylinder wall. Circle-track gas ports should be lateral, not vertical.

Piston Rings

Ring seal is a key issue to optimizing engine performance, regardless of the parts used. The most basic race engines will benefit from attention to detail in this area. It is the job of the ring pack to keep the cylinder pressure from bleeding into the oil pan and to keep oil from getting into the combustion space.

Reducing ring tension can lead to significant power gains from friction reduction, but only if ring seal is not compromised. Running smaller ring widths, such as 0.43 inch, 0.43 inch, 3mm (as a stack), will help reduce loss, but the margin for error in the remainder of the cylinder kit is reduced. In particular, top ring land flatness becomes more critical and bore finish must be absolutely correct.

Smaller rings have less area to transfer heat and tend to run at a higher temperature. Such increases in temperature can lead to micro welding. Small rings usually will not have the longevity of a larger ring, due to the reduction in contact area and increased operating temperature.

Producing a vacuum in the sump, either from a dry-sump system or an external vacuum pump (where legal), can dramatically increase ring seal and allow for a more aggressive ring pack.

Piston Ring Materials

Most rings today are made from iron. The ring face has a channel or groove cut into it with a sprayed-in moly filler. The ring is then finish-ground. This combination is very forgiving and works well on most cylinder bore surfaces.

Steel rings with Physical Vapor Deposition (PVD), Titanium Nitride (TiN), or ceramic coatings are being used successfully in other forms of motorsports. These rings and coatings are hypersensitive to the type of bore material and finish required. In some instances, these coatings may offer some advantages with a Nikasil-coated bore. Such combinations may yield the lowest friction yet but can be extremely difficult to achieve ring seal.

Measuring Piston Ring Tension

The traditional way to measure ring tension is to load (install) the ring into the bore and use a spring scale to determine the force required to pull the ring. This is a crude but effective way of measuring ring tension, more commonly called ring "drag." Properly performed, this method requires a great deal of engine builder experience to be truly effective.

OEMs measure ring tension with a machine that loads the ring tangentially. This type of machine is costly but is extremely accurate and repeatable. A tangential ring tension gauge usually loads the ring into a dummy ring grove while a steel band is wrapped around it. Bearings allow the band to apply a true tangential load to the ring. One end of the band is attached to a load cell to measure force. The other is pulled. Load is applied until the proper ring gap is found and the force is read. Eventually, as the practice of low-tension rings continues, more engine builders will use this type of gauge. It will be the only way to effectively measure the ring tension, with ultra low-tension rings.

CYLINDER PRESSURE VS. CRANK ANGLE

Study the illustration on page 8 showing cylinder pressure history as a function of crankshaft angle. Note that peak cylinder pressure during "normal" combustion occurs just past TDC. While piston ring seal is important during the compression stroke and prior to ignition, it is especially critical to create and maintain pressure seal in upper portions of the cylinder in this post-TDC area. The higher the rpm, the longer the delay before peak pressure is developed.

The measurement of engine output often includes a comparison of negative and positive work performed on a piston. Simply stated, pre-combustion pressure ("work") is applied to a piston on its upward stroke during compression. Depending upon the crankshaft angle at which ignition occurs (prior to TDC), this amount of "negative" work will vary. In other words, the earlier the spark, the greater the negative work applied, all else being equal. In fact, again noting the illustration, the nearer the point of ignition (A) and the earlier the point of peak cylinder pressure (P1 or P2), the greater the amount of position work applied to the piston. Once a piston has passed TDC on its power stroke, positive work is applied.

At the risk of oversimplification, the difference between positive and negative work is a measure of IMEP (indicated mean effective pressure). As an aside, the faster the combustion rate (burn), the later can be the initiation of spark—resulting in decreased negative work and increased positive work on the piston. Technically, indicated work is the difference between work done on a piston during its compression stroke and work done on it during expansion. Graphically, this relationship is often expressed in the form of P-V (pressure-volume) diagrams or so-called pressure "loops."

Based on decreasing combustion space and rising combustion temperature, controlled cylinder pressure (absent any pre-ignition or detonation) will peak past TDC, owing in part to delays from gas dynamics. Ordinarily, it is within the first 30 crankshaft degrees past TDC where ring seal is particularly critical. This leads to the necessity (particularly at this point) for pistons being square in their bores, rings that are in proper contact with cylinder walls, bore concentricity at temperature, and skirt drag minimized for friction horsepower loss reduction.

Chapter 3

RACE ENGINE BASICS: CYLINDER HEADS, VALVETRAIN & INTAKE MANIFOLDS

by Jim McFarland & Chris Jenckes

Three important areas for engine performance improvement are the cylinder heads, valvetrain, and intake manifold. Remember, heads that flow the most are not necessarily the best for your application.

From Street Stock to Nextel Cup, three of the most significant and popular areas for engine performance improvements are cylinder heads, valvetrains, and intake manifolds. In particular, cylinder heads affect most aspects of power. Ports, valve size, and placement affect engine breathing or volumetric efficiency. Thermodynamic and combustion efficiencies are affected by the volume and shape of the combustion chamber. Valvetrain performance is affected by placement of valves relative to the rockers. The water jacket design can simultaneously affect cooling performance, power, and overall engine durability.

Cylinder heads that flow the most air on the flow bench may not make the most power or perform the best on the track. Cylinder head flow gains are not an end unto themselves. For circle-track racing, a primary goal is to pack as much air and fuel into the cylinder as possible while maintaining the highest port velocities. This will not necessarily occur with a head that flows the most air on a bench. We'll call this "transient" airflow or the rate at which a given cylinder head design can deliver air.

PROCESS OF EVOLUTION

Upon evolution of an economical airflow stand or bench, engine builders were quick to see the tool was an effective way to quantify airflow performance of one port relative to another. The flow bench is most effective when used to develop cylinder heads that are closely related to production parts. The reason is that airflow performance of a production part is so restricted that any gains in flow will most likely increase engine power.

Most flow bench testing is performed at a depression of 28 inches of water. At best, this is accurate only twice during a full inlet cycle (See Figure 1). This chart shows an inlet port that sees negative pressures up to 2.6 psi (73 inches of water), much higher than the 28 inches of water used on an airflow bench. A true representation of cylinder head flow performance would use a three-dimensional plot of flow at varying lifts and depressions. Flow benches become less effective as cylinder head design is less restricted and mass flow increases. For example, heads developed for CART, IRL, Pro Stock drag racing, or Outlaw sprinter engines are less likely to be successfully developed on a standard airflow bench.

Further, an intake port must flow fuel and air. Engines running methanol operate at air/fuel ratios more than twice the richness of those on gasoline. The large quantity of fuel, much heavier than air, clearly impacts the performance of an intake port. Wet-flow benches that can operate at much higher flow depressions make a better tool for such engines. Some engine builders working on these engines simply use the dynamometer to test cylinder head improvements. The flow bench becomes a quality control tool, much like a

Figure 1 As referenced in the text, this chart compares a typical trace for an intake valve's motion and the multiple excursions of pressure ("spikes") that occur in an intake path (cylinder head and intake manifold runner) prior to and during an inlet cycle. Note not only range of pressure (minimum and maximum values) but also nature of pressure history during intake cycle (crank angle from about 300 to 620 degrees as shown).

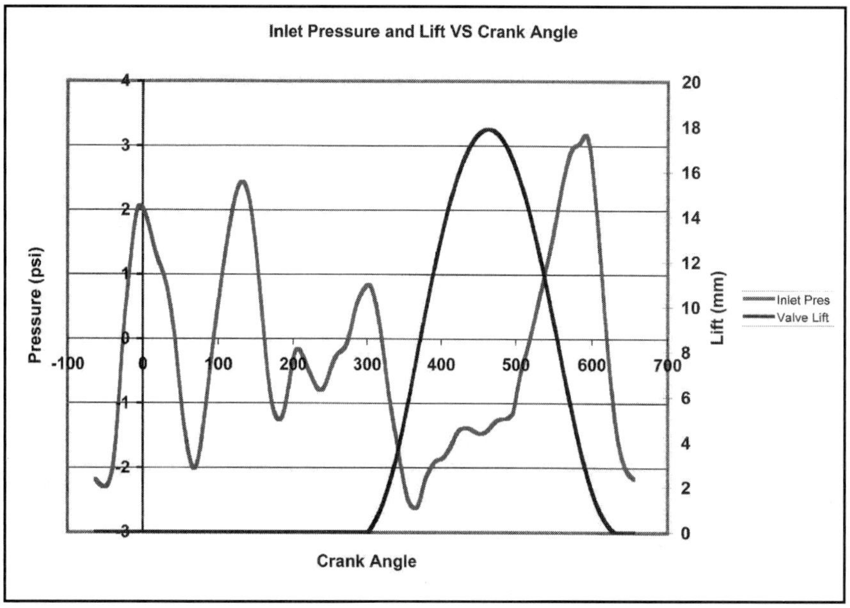

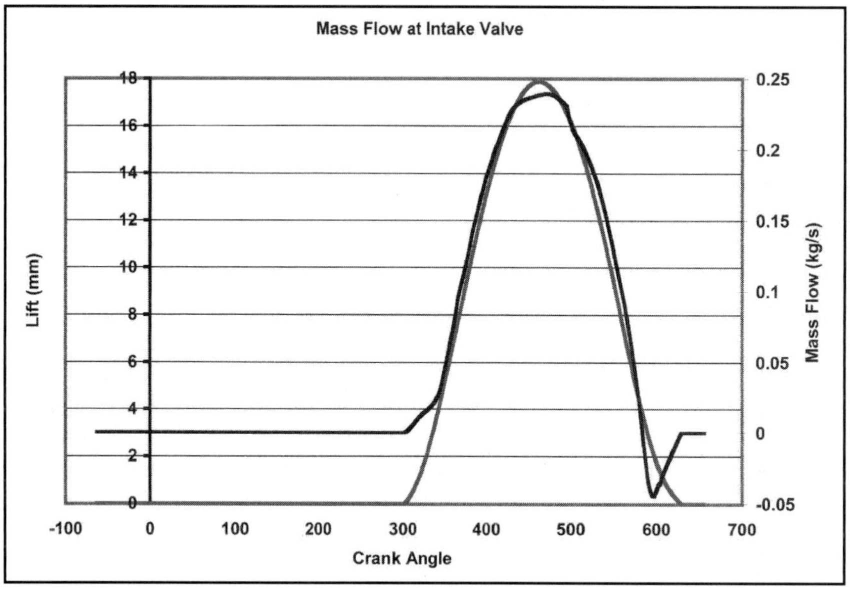

Figure 2. This illustration compares a typical intake valve motion to mass flow during the filling event. Particularly note reduction in volumetric efficiency around peak valve lift, suggesting flow gains in this area are of lesser significance than at low-lift values.

go/no-go gauge, to verify that a finished head compares to its prototype. The point here is that the Saturday-night racer should not become overly dependent on airflow benches. The dynamometer is a superior engine evaluation tool.

ON THE INTAKE SIDE

Where in the valve event does flow take place? On the chart, note intake mass flow versus crankshaft angle. In this case, one area on which to work is at low valve lifts. Since the piston has started moving downward on the intake stroke, there is already negative cylinder pressure as the intake valve opens, causing a jump in mass flow. Note that the rate of mass flow increase flattens at high lifts, so increasing valve lift does not appear to be a good option for more flow. Also note that some reversion takes place. (This example was taken from an engine speed near peak torque.) A compromise in the valve events must occur to have good power at higher engine speeds. The reversion will disappear at power peak.

So, how do you improve the cylinder head? It's possible to improve performance of the basic head, even if you don't have a flow bench. The easiest way to improve low-lift flow is to work in the valve seat area. Blending from the seat to the bowl (without damaging the seat) will increase low-lift flow. Using a radius-type valve seat cutter with a live pilot can produce better flow than traditional types of valve seat machines. If you're having a shop cut seats for you, find personnel with experience in this type of equipment.

Back-cutting valves is usually another low-lift flow gain method. If you have access to a valve seat re-facer, start with a 30-degree back-cut for Street Stock-type applications. Position the back-cut so it narrows the 45-degree seat to match the valve seat's width.

In addition, most production cylinder heads have excess material around the valve guide due to casting (draft) requirements. Faring the guide boss into an airfoil cross section and removing material in the bowl area immediately adjacent to the guide is usually productive. Go slowly when removing material.

Along the intake and exhaust paths, developing and maintaining optimum flow energy (momentum) is important to net volumetric efficiency. This is an important consideration, and here's the reason along with some food for thought. It takes energy to accelerate air. It also requires a

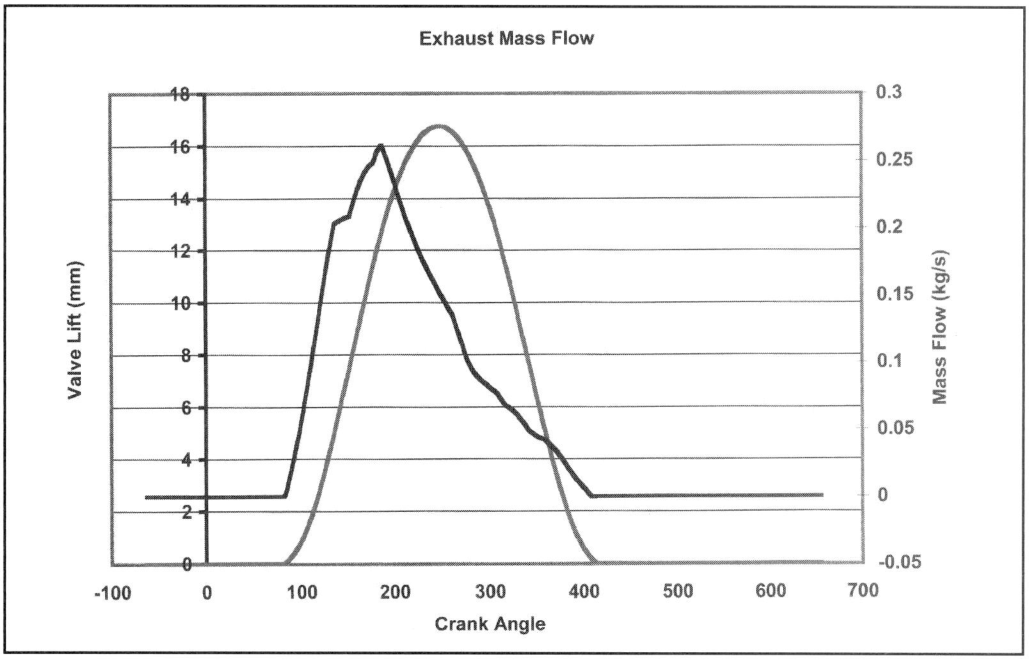

Figure 3. Largely because of blow-down pressure (residual combustion pressure at the time of exhaust valve opening), this illustration shows a period of a typical exhaust valve cycle when exhaust flow is the greatest. Compared to the closing side of an intake valve event, note how little flow occurs during exhaust closing.

given amount of energy to move air along a constant cross-section duct. If the duct (passage) expands, the air will slow. If the duct then contracts after the expansion, the air will accelerate. This acceleration robs the air stream of energy. On the intake side, this alternatively changing rate of flow can hamper the ability of air and fuel to stay mixed (homogeneity). Many novice porters make the head opening larger on the pushrod side of the port. This allows the intake charge to expand and then contract to clear the pushrod, exacerbating the problem.

Remember, the goal of intake port development is to fill the cylinder as efficiently as possible. Energy wasted in reaccelerating the intake charge after a duct area expansion is not helping to fill the cylinder. This issue also speaks to the subject of maintaining passages of fairly equal cross section, or at least those that don't vary in much area along a given flow path.

WHAT IS MOMENTUM?

Simplified, momentum is a measure of energy. Mathematically, it is a product of a body (or system) mass and linear velocity. In an inlet path, we're dealing with a system (air and fuel) of different mass trying to move at the same velocity. Changes in either mass or flow rate will change this system's momentum, thereby causing a companion change in the system's energy level.

More simply stated, if air/fuel charges moving along an inlet path experience a change in momentum (energy) by alternatively accelerating or decelerating their velocity (by companion changes in cross-section area), either of two primary conditions may occur: (1) Air and fuel may mechanically separate, and/or (2) net energy may be lost to optimizing the filling of the cylinder or volumetric efficiency. Both are undesirable.

While it may not be possible to maintain zero changes in momentum along an inlet path, particularly since we're dealing with bi-directional and unsteady flow, keeping these changes to a minimum will help boost the filling of the cylinder and net torque.

ON THE EXHAUST SIDE

In similar fashion, the exhaust side of a cylinder head responds much like the intake. For production-based engines, the exhaust valve responds well to a radius from the face to the edge. Back-cuts should also be tried. Work the bowl and guide areas similar to the intake. A generous radius from the seat to the short turn is best, and try not to remove metal from the floor of most production heads.

REVERSE FLOW

Many anti-reversion devices have been tried to keep the direction of the charge (or flow) constant. This theory would be interesting if the anti-reversion device could be timed to prevent reverse flow only when it was undesirable. (Assuming reverse flow is sometimes desirable.)

We do know that scavenging and sonic tuning must have an effect since racing engines routinely have volumetric efficiency in excess of 100 percent. If scavenging or sonic tuning does occur, then the flow is pulsing in the intake and exhaust tract and going in both directions (not simultaneously) for benefit. If reverse flow can assist scavenging and sonic tuning, then the reverse flow of the port must be of some significance.

In most cases, it turns out that modifications

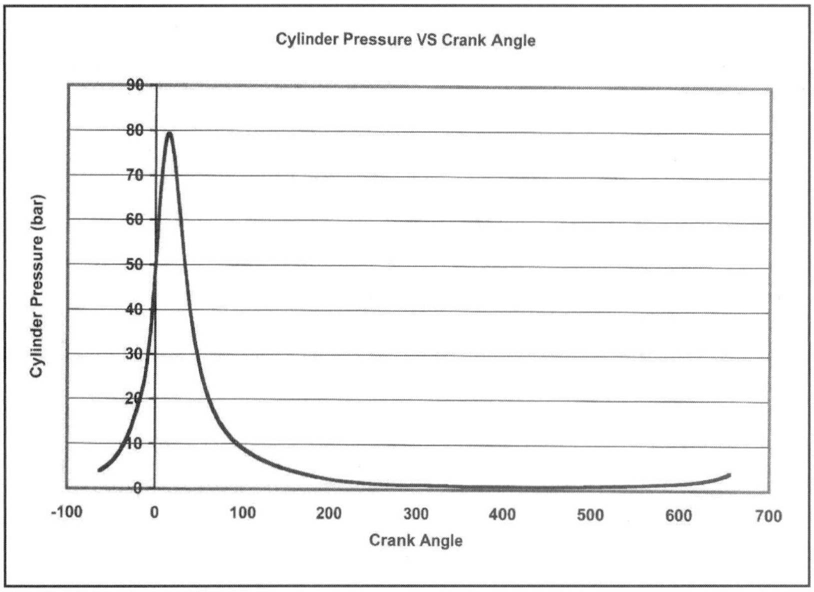

Figure 4. Note here that peak cylinder pressure occurs past TDC and that this point is developed further past TDC as engine speed increases . . . something to keep in mind when either tailoring ignition spark curves, setting initial timing, or both.

improving forward flow also improve reverse flow, and it is not a good idea to sacrifice forward flow for reverse flow. Nevertheless, it is good practice to flow ports in both directions. Any modification that can be made to improve reverse without affecting forward flow is a good idea, but tuning changes may be required for optimization of any change.

WHAT ARE OVERLAP AND SCAVENGING?

For a variety of reasons, including improved volumetric efficiency (VE) at high rpm, camshafts may be designed to delay exhaust valve closing to overlap when its corresponding intake valve begins to open. During such time, incoming air/fuel charges may pass directly into a cylinder and out through its exhaust. In one sense, fresh air/fuel charges may be lost from combustion. In another, unburnable exhaust gas from the preceding firing cycle may be removed from the cylinder. If we define scavenging as the process enabled by intake-and-exhaust event overlap, it's possible to examine how to achieve volumetric efficiency greater than 100 percent.

Over time, numerous terms have been used to describe this process. Whether called sonic tuning, wave tuning, ram tuning, or some other applicable term, the objective is to achieve a greater volume of air/fuel charge filling a cylinder at the time of ignition than could be achieved under the singular influence of atmospheric pressure. While the term scavenging often applies to two-stroke-cycle engines (particularly scavenging ratio and scavenging efficiency), the ability of a four-stroke-cycle engine to utilize some form of pressure excursions/dynamics to enhance VE involves valve timing and dimensional properties of the intake and exhaust systems.

Keep in mind that the point at which the intake valve opens is critical to how much residual cylinder pressure (and exhaust gas) is present. As opposed to the suggested benefits of bi-directional flow in the passage tuning process, this initial reverse flow or reversion pulse can dilute fresh air/fuel charges subsequently flowing into the cylinder. Exhaust flow efficiency also affects the residual volume and pressure of unburnable gas available to the intake system at inlet valve opening. It is beyond this initial point that bi-directional flow may produce benefits to the tuning process. Ideally, the scavenging process would enable replacement of exhaust gases with fresh air/fuel charges without any loss of such charges to the exhaust system. In reality, due to multiple and highly complex factors, optimum scavenging needs to be "timed" with engine speed closely associated to the rpm most often used. Even the most comprehensive of contemporary computer modeling systems require final analysis on an engine dynamometer (often of inertia type) and on-track evaluation. Specifics of the issue are beyond the scope of this particular story, but it is fundamental to camshaft, intake, and exhaust system parts selection that components be chosen that perform in a predetermined range of engine speed (and as a package), or the benefits of both overlap and scavenging will not be achieved.

THE MULTIFUNCTIONS OF CYLINDER HEADS

Too often, cylinder heads are viewed only for their ability to help achieve high or increased airflow. In fact, although these parts can contribute significantly to total airflow and specific flow patterns, it is in the combustion space that any quantities of air (or air/fuel charges) must be converted into useful heat. So in the design or modification of intake and exhaust ports and combustion chambers, it is particularly important to note how inlet air is being delivered to the combustion space.

Various terms have been used to describe the quality of inlet air. There will be certain airflow characteristics dictated by the relationship between the intake port flow path and the cylinder bore. For example, an inlet port may be positioned or modified to either increase or decrease the circular and downward motion of flow as it enters a given cylinder bore, somewhat like twisting or screwing the air column into the cylinder.

If an intake port is created that forces air/fuel charges against the inside diameter (ID) of a cylinder bore, mechanical separation of air and fuel may occur, leading to reduced mixture quality and combustion efficiency. In a sense, intake and exhaust ports need to enhance flow into and out of the combustion space, not just show substantial quantities of flow on an air bench.

Top motorsports teams often utilize computer simulations of air movement along an inlet path and into the combustion space. In the design or investigation stage, multidimensional computer programs such as Computational Fluid Dynamics (CFD) can be used to construct three-dimensional models enabling rapid analysis of various changes to portions or functions of an engine. Even though some of this technology is still in a developmental state, sufficient precision exists to include this method in developing cylinder heads, intake manifolds, and related components. What this says to the Saturday-night engine builder is there is more to airflow than quantity. In *Circle Track* tech articles, the importance of airflow quality has been discussed and stressed. Beyond the benefits of CNC-machining processes and the ability to duplicate ports and combustion chambers designed or modified by hand, it's important to determine port and combustion shapes optimal to combustion efficiency and power before applying the techniques of CNC-machining.

As crude as the process may appear on the surface, the use of spray patterns during air bench port and chamber analysis and/or modification remains a powerful tool to the Saturday-night engine builder seeking to derive some of the benefits of CFD from a "toolbox" or non-computerized approach. In the process of using this method, creating ports whose airflow quality helps reduce air/fuel mixture separation in the combustion space (as evidenced by dye patterns showing streaks of liquid fuel on chamber or cylinder walls) while still optimizing net flow will typically lead to solid power gains.

VALVETRAIN CONSIDERATIONS

Valve un-shrouding is one of the easiest ways to improve airflow performance of a cylinder head. The simplest way to do this is to place the head on as big a bore as practical for your particular engine. It is safe to say the more a valve is lifted from its seat, the greater the distance it should be from the chamber and cylinder walls. Canted valve heads and hemispherical heads naturally have this feature. Typically, the intake side is more sensitive than the exhaust side and the long side is more important than the short side. There are physical limitations to what can be done, but compromising the combustion performance of the chamber for greater flow is not an acceptable trade-off.

In Nextel Cup racing, a key to improving engine performance is to improve performance of the valvetrain. A significant portion of Cup team engine development is spent working with valvetrain spin fixtures. These devices allow gross valve motion to be measured at different simulated engine speeds.

The results show valve timing and lift effectively change with engine speed. Durations measured at 0.050 and 0.100 inch regularly vary 10–15 degrees from low to high rpm. Loft (additional lift from lobe-follower separation at maximum lift) can exceed 0.080 inch on some roller profiles.

While most racers don't have access to a spin fixture, it is clear that comparing camshafts by looking at 0.050-inch duration numbers and lift is like comparing cylinder heads by valve size. What the Saturday-night racer should do is compare cams at multiple durations, such as 0.004, 0.020, 0.050, 0.100, 0.150, 0.200, and 0.25 inch. This type of comparison will produce a better understanding of how the lobe is likely to perform. What should you look for? Modern cam profiles have significantly higher low-lift acceleration rates. This means the duration on the seat and at 0.050 inch will actually be smaller, but the lobe will have the same duration at 0.250 inch. This type of profile allows you to use a smaller cam for torque but still make power where expected.

Talk to the cam manufacturer and ask about the dynamic performance of the lobe (i.e., its limit speed). The limit speed is the speed beyond which the valvetrain cannot be turned any faster. This should be much greater than the maximum intended engine speed. Find out the engine speed at which significant bounce occurs. This speed should just exceed maximum anticipated engine speed. It is important to know exactly what components were used during the manufacturer's spin fixture analysis. If your valvetrain varies from what was tested, the data may not relate to your particular engine.

How do components affect dynamic operation of the valvetrain? In general, you want to remove as much mass as possible from any part on the valve side of the rocker fulcrum (valve, retainer, and locks). This allows the spring to do a better job, helps reduce bounce, and increases limit speed. On

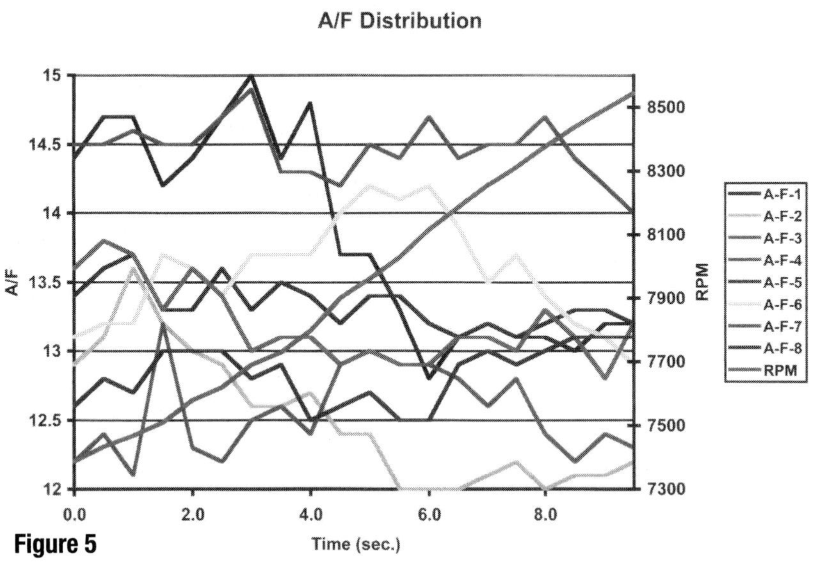

Figure 5

the lifter side of the fulcrum, stiffness becomes more important than mass (weight). A stiffer pushrod will follow the lobe better. Reducing the mass of the lifter and/or pushrod will reduce loft; increasing mass will increase loft.

Most of us don't have access to a spin fixture, so work with your cam manufacturer to select the correct valvetrain components. Remember, the valvetrain is a system, and changing any component can adversely affect the performance of that system.

For serious circle-track engines, valvesprings are life. A better spring will allow a more aggressive valvetrain and, in turn, will make more power. Always use the minimum spring pressure that can be safely run. More spring pressure than required will generate heat and rob power. Keep track of the installed pressure of springs before and after every race. Replace all the springs at the first sign of a significant decrease in pressure.

Treat valvesprings with care. Don't nick or scratch them because this may affect their durability. In high rpm operations, valvesprings generate tremendous heat. If spring durability becomes an issue, valvespring oilers can help but will increase oil flow that can cost power.

DEEPER INTO THE VALVETRAIN

Knowing the basics is not sufficient. Equally important is understanding what areas of power development are affected by the valve function. Without this knowledge, power may be lost in the quest for simply doing what's necessary to prolong valvetrain life. Depending upon the class of engine, the length of time between planned valvetrain maintenance (particularly valvespring inspection and replacement) may range from after every race to when something fails. Regardless, the following information applies.

By their motion, valvesprings generate heat. It is important that they receive adequate oil for cooling. Also, in the process of being compressed and extended, they never come to rest. Even when valves are seated, residual harmonics, or compression and extension of spring coils within a given stack, continue between times when valves are opening or closing. These movements are not always along the axis of a given spring, resulting in lateral distortion that accompanies axial motion. This produces components of spring motion that are counter to what might be called pure spring compression and extension.

Despite the efforts of camshaft designers to provide stable valve motion during low-lift opening and closing points, there remains a system of flexible parts between the valve tip and cam lobe (pushrods, springs, and rockers). These components contribute to valve motion that is not true to the cam's lobe profiles, thereby producing lift patterns less than what is designed into a given camshaft. During valve seating, reducing the elastic collision between valve heads and seats is an ongoing problem. Engine builders can help the situation by applying manufacturer-recommended spring installation and pressures throughout the life of a given spring.

Springs should be sufficiently stiff to control valve motion approaching maximum lift, as well as during its acceleration to closing. As engine speed is increased, the more important spring forces required to control valve motion occur during maximum valve acceleration. Boiled down to its essentials, this points to the importance of maintaining sufficient spring pressure just prior to maximum valve lift in order to maintain lobe/follower contact at and just beyond peak lift. Even though multiple springs can add net pressure to a given spring package, this approach is also important to controlling surge by combining different natural frequencies of individual springs. Spring design that includes so-called beehive or tapered shapes can also provide desirable harmonic damping.

MANIFOLDS

It is appropriate that the intake manifold is mounted on top of the engine. After valve events have been selected, the intake is the primary tuning device for a four-stroke spark ignition engine, just as the expansion chamber is for a two-stroke. In the case of a carbureted V-8, the intake manifold recognizes incoming air/fuel charges and directs them into the cylinder head(s). Tuning becomes the manifold's second function.

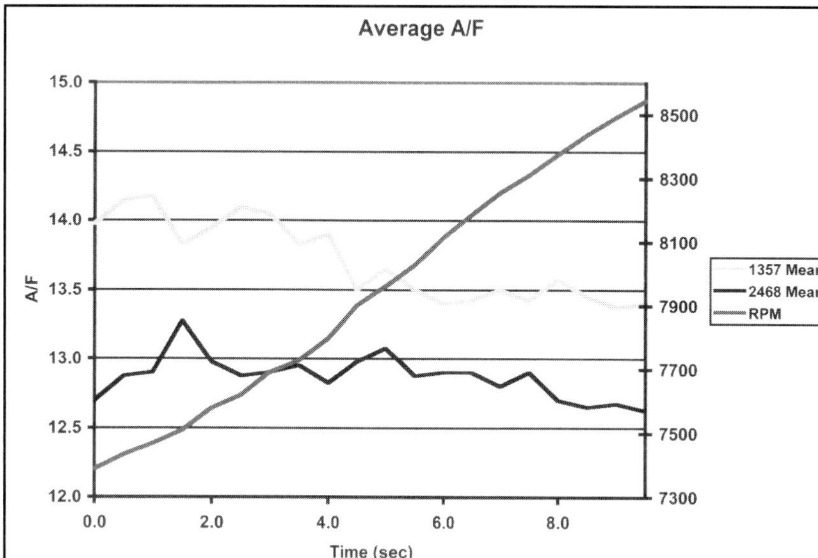

Figure 6

AIR/FUEL CHARGE DISTRIBUTION

While there have been many magazine articles written on how intake manifolds can be used as a tuning component and how to select the appropriate manifold for your engine, few discuss the variation of air/fuel ratio from cylinder to cylinder. This is a critical factor in engine tuning because the mixture can only be leaned to the point where the leanest cylinder is at its operational limit, particularly with single-point fuel delivery, as in the case of a 4V-carbureted engine. With individual runner (IR) induction systems and electronic fuel injection, this cylinder-to-cylinder variation can be tuned to less than 0.5 air/fuel ratio. Carbureted V-8 engines typically have significantly worse variations that can be up to four air/fuel ratios from the worst cylinder to best.

Cornering g-forces can have significant effect on overall mixture distribution. This can be seen when comparing dynamometer air/fuel ratio data to on-track data. Figure 5 shows on-track air/fuel ratio data from a GM ARCA engine at a 1.5-mile high-speed track. All four carburetor main jets were identical. The data shows a variation from 12.0:1 for cylinder No. 2 at 8,100 rpm to 15.0:1 for cylinder No. 1 at 7,700 rpm. For optimum power, three air/fuel ratio variations from cylinder-to-cylinder is an undesirable condition. When tested on the dynamometer, this manifold showed a 2.0–2.5 air/fuel ratio variation, verifying the inherent differences among ratio spread on an engine dynamometer and at the track.

Figure 6 compares air/fuel ratio averages from the left bank (cylinders 1,3,5,7) to the right bank (2,4,6,8) of a V-8 engine with firing order 1-8-4-3-6-5-7-2. As you might expect, the data shows the impact of the g-forces makes the right bank richer than the left. This effect will be more pronounced on tracks with high cornering loads and can be minimized by stagger-jetting the carburetor. This example was selected because it clearly demonstrates the point. Not all manifolds are affected this severely.

IMPROVING CYLINDER-TO-CYLINDER DISTRIBUTION

If an air/fuel distribution problem exists, make certain the carburetor is placed properly on the intake. Another method of adjusting distribution is to move or bend the carburetor booster relative to the throat in which it's installed. (Exercise extreme care when attempting to bend boosters. It is also possible to place small tabs or protrusions on the exterior of booster bodies to redirect airflow in the throat and alter post-carburetor mixture flow direction.) Carburetor spacers may also have an impact on distribution (see section discussing spacers, page 23). Often a four-hole or "combination" four-hole and open spacer will improve cylinder-to-cylinder distribution.

At best, working with the manifold itself to improve distribution is difficult and should only be attempted when a dynamometer is available with eight channels of air/fuel ratio sensors, one for each cylinder. Reading spark plugs may not be accurate enough for this type of development.

The most significant factor that affects cylinder-to-cylinder air/fuel distribution in the intake manifold is the spatial relationship between the runner openings into the plenum and the carburetor flange. The floor of the runner is usually the most sensitive area. Adjusting the runner opening to more of the plenum will generally make that cylinder richer. Often the center cylinders shroud the end cylinders in a V-type engine. Careful unshrouding of the end cylinders can richen a lean-running cylinder.

MANIFOLD SELECTION

Typically, the selection of manifold configuration is limited for circle-track applications. The choices are single-plane or dual-plane. Usually, engine

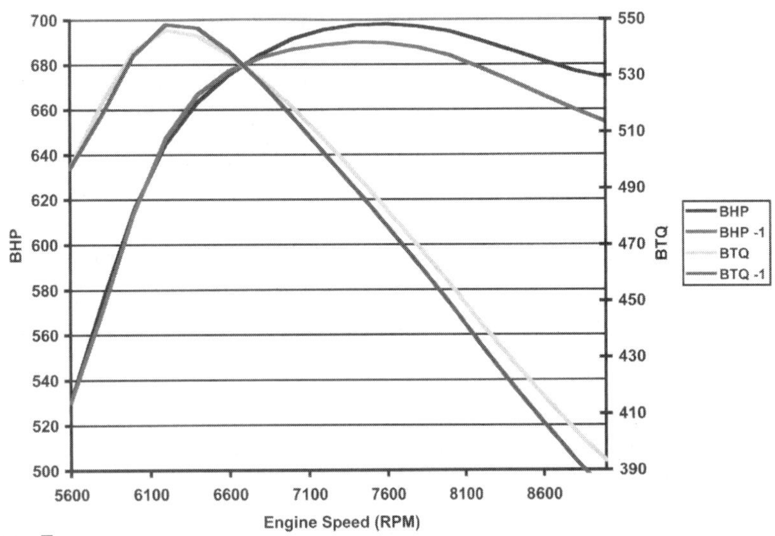

Figure 7

operation speed determines the manifold configuration. If the power peak of the engine is below 6,500 rpm, then a dual-plane is most probably the manifold of choice. As the power peak moves significantly above 6,500 rpm, the choice moves to a single-plane. The choice becomes more complicated when the engine's operation speed is somewhere in between. In this case, there is no clear answer, so testing on the engine is necessary. If well-developed (modified to suit a particular engine package), a dual-plane can be more responsive and may be the driver's choice.

DON'T DISCOUNT THE MULTIPLE INFLUENCES OF INTAKE MANIFOLDS

Intake manifolds can make or break a good set of cylinder heads or be easily flawed by improper use or modification. Regardless of which design or brand chosen or changes made, there are some essentials to keep in mind.

Intake manifolds don't flow in one direction. There are times, depending upon engine speed and load, when pulses are directed back toward the carburetor (or air inlet point). Technically, this describes bi-directional, unsteady state flow. Despite the way this is labeled, these reverse-flow pulses are disruptive to overall airflow and air/fuel mixture quality (homogeneity). Either condition can impair power. Therefore, there are only certain intake manifold features that can be evaluated on an airflow bench, although these include mapping of flow pressure profiles (pressure distributions) and specific velocity patterns.

It's also important to recognize intake manifold pressure conditions that encourage increases in unburnable combustion residue (principally exhaust gas). For example, conditions creating some level of manifold vacuum at wide-open throttle allow more contamination of fresh air/fuel charges than the near-zero vacuum when the influence of atmospheric pressure is greatest. This can be found in either restricted engines or those required to run two-valve or small four-valve carburetors.

In these instances, intake manifolds or carburetor spacers designed or modified to thwart reverse flow help minimize diluted mixtures and improve power. Another consideration is camshafts and exhaust ports/valves that address reverse exhaust flow as further dampers on reversion, particularly concerning exhaust opening timing. These determinations can be made on an air bench by flowing ports in the reverse direction, particularly at low valve lifts.

Overall, the problem of changing the direction of air and fuel flow from approximately vertical to the entry angles of intake manifold runners is crucial to delivering efficiently combustible mixture. Air tends to respond more quickly to throttle changes than fuel. Air and fuel are also prone to separate. In this sense, post-carburetor air/fuel mixture preparation becomes a function of the intake manifold. Surface finish in a manifold's interior can also play a role, trending toward rough instead of smooth surfaces in order to help create or maintain efficient atomization of post-carburetor fuel.

Although it may be difficult to separate the need for mixture quality from net airflow, each must be considered vital to proper manifold function. If Cup engine builders didn't acknowledge this as a key to optimizing power, the inordinate amount of time and expense devoted to intake manifold and carburetor spacer (or restrictor) applications would not occur. While the Saturday-night engine builder may not have the resources to address these problems similarly, skillful use of airflow benches can be a suitable substitute. It's important to become familiar with various air-bench techniques that extend beyond mere mass flow measurements to include pressure patterns and airflow quality.

Like many engine components, intake manifold selection (as rules permit) should include specific ranges of engine speed most frequently used. While peak power numbers may be impressive or applicable in certain situations, torque production within an intended span of rpm is important to overall race car performance.

Rod length also plays a role. As piston speed around TDC is decreased (with rod length increased), it's helpful to use intake manifolds and inlet port sizes trending toward smaller section areas that aid flow velocity, independent of large piston displacements and high rpm. The rate of pressure drop across the inlet path (boosted by smaller runners) aids volumetric efficiency in the lower and middle rpm. In fact, it is worthwhile to consider an intake manifold's runners as extensions of intake ports, requiring mutual compatibility in airflow potential and uniformity of pressure distribution. The latter is particularly important at the interface between manifold and head surfaces. Manifolds that neither reduce port flow nor (by themselves) exceed port flow can be considered extensions of a cylinder head.

Be assured—investigations by Cup engine builders include some, all, or more of these areas that influence the overall intake manifold function. By the same token, the Saturday-night engine builder can choose and apply those that seem to fit particular power or on-track requirements.

SMALLEST VOLUME POSSIBLE

In circle-track racing, it is generally best to approach manifold development or modification by starting with the smallest overall manifold volume possible and increasing volume until no power gain is observed. By having the smallest volume possible, the engine is more responsive to throttle changes and generally easier to drive. By their nature, larger volume manifolds respond less quickly but may perform better at higher engine speeds.

TUNING INTAKE MANIFOLDS

When modifying intake manifolds, the three most common tuning elements are runner length, runner taper, and plenum volume. Using an Engine Simulation Program (ESP), the graphics provided in this section were created by modeling an ARCA engine. ESPs are one-dimensional models that perform wave calculations and are useful in the design of intake and exhaust systems. These high-end computer programs can predict the volumetric efficiency of an engine to within 2 percent of actual run data.

RUNNER LENGTH

Runner length tunes an intake manifold based upon pressure waves, or sound. The longer the runner, the lower the engine speed range in which the tuning will take place. To illustrate this concept, two computer models were prepared: a baseline model and a model with 1 inch of length added to all the runners and no other changes made.

Figure 7 (page 20) shows the power and torque curves of the engine with the two manifolds. Clearly, the addition of 1 inch of runner length increases peak torque and moves the power peak down 200 rpm. Peak torque increased 2.0 lb-ft, but peak power was down 7.9 bhp (brake horsepower). At 7,600 rpm, the power of the longer manifold is 8.3 bhp less than the baseline. While these specific quantities may not be significant, their direction validates the theory behind the change.

Figure 8 (above) shows the pressure versus crank

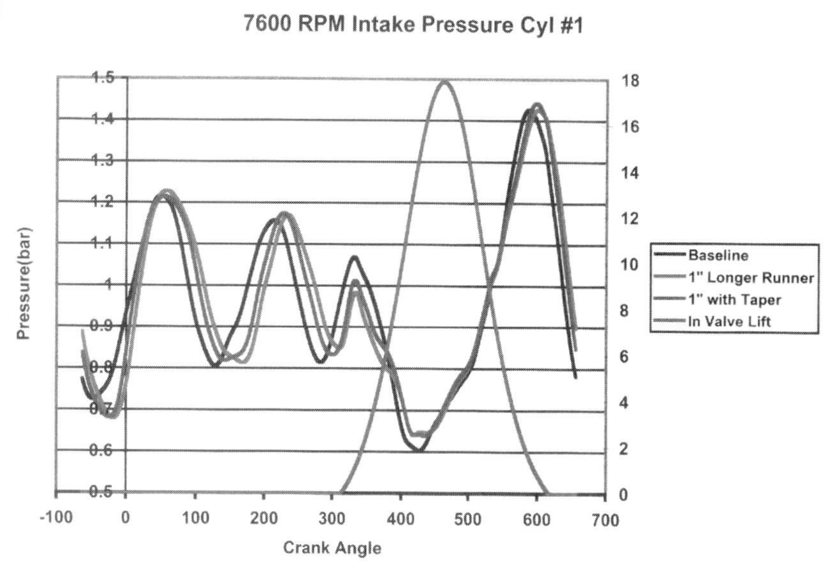

Figure 8

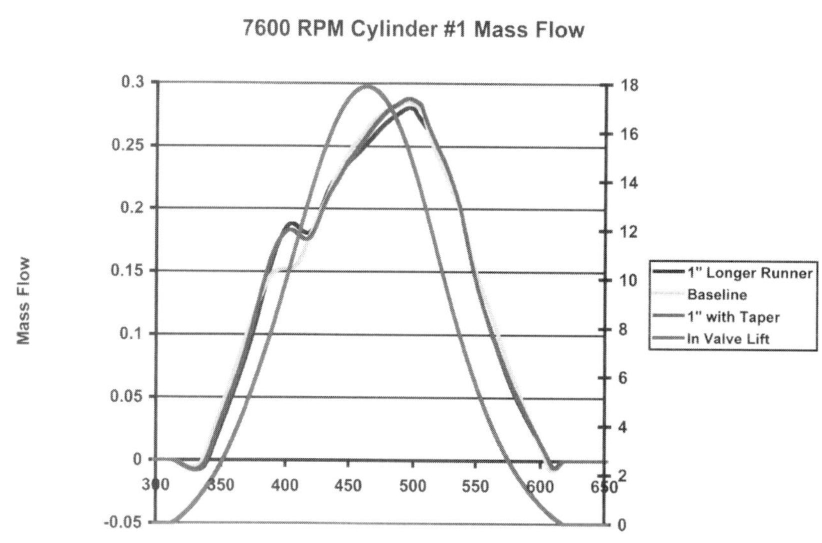

Figure 9

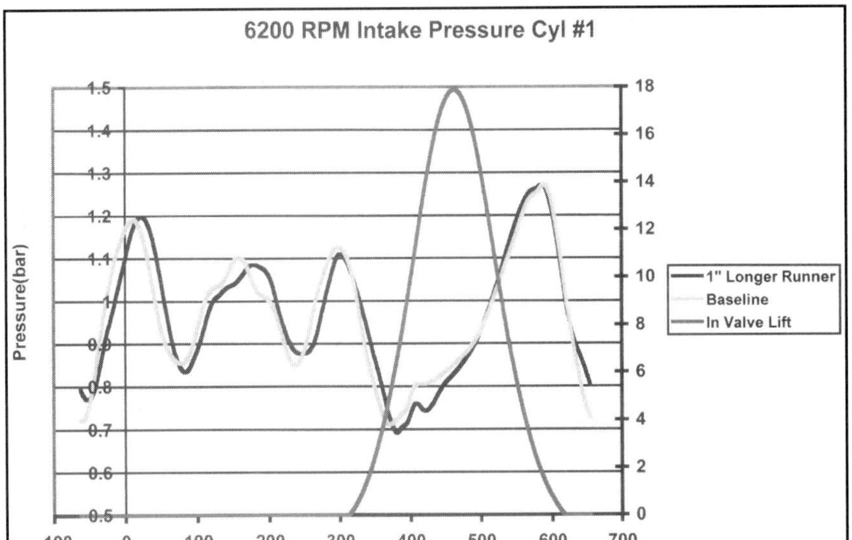

Figure 10

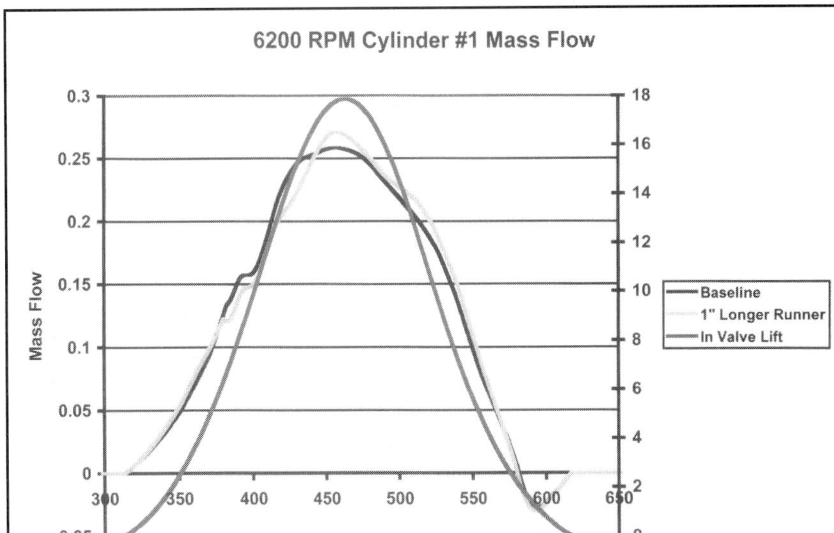

Figure 11

angle at the outlet of the intake manifold (cylinder head juncture) at 7,600 rpm, power peak for the baseline. An increase in pressure here means that the charge density is higher and a better cylinder fill will occur (higher volumetric efficiency). The baseline manifold is doing a better job shortly after the valve opens until just before it closes. It is interesting to note that both cases had volumetric efficiencies above 100 percent. The pressure versus crank angle data shows how this can occur. When the intake valve closes, there is significant pressure above ambient (pressure) in the port that provides a mild supercharging effect. (Note: One bar is 14.7 psi or the equivalent of atmospheric pressure or 1 atmosphere.)

Overall, filling the cylinder is the goal. More mass flow into the cylinder means more power. Figure 9 (page 21) shows mass flow versus crank angle at the intake valve at 7,600 rpm. The greater the area under the mass flow versus crank angle curve, the more chemical energy is available for power production. The shorter runner manifold traps more mass in the cylinder at this engine speed than the longer manifold, making more torque and power.

Figure 10 shows the pressure versus crank angle at the outlet of the intake manifold at 6,200 rpm (torque peak). The longer runner is doing a better job from peak lift until the intake valve closes. Again, there is pressure above ambient (atmosphere) at the close of the intake valve in both cases.

Regarding cylinder filling efficiency and manifold runner length, Figure 11 shows mass flow versus crank angle at the intake valve at 6,200 rpm. The longer manifold runner traps more mass in the cylinder at this engine speed than does the baseline.

RUNNER TAPER

Taper is the relationship between the size of the runner's inlet opening and the size of the same runner's exit. The effect of increasing taper (the opening is larger than the exit) is to foreshorten the runner. The greater the taper, the greater the foreshortening effect. Taper is especially useful when it is difficult to shorten the runner length.

In the previous intake model, the 1-inch-longer runners were modified by adding a significant amount of taper to all of the runners to see if the power could be recovered. The bhp improved by 11.6 over the longer runner alone and was up 3.3 over the baseline. Again, reviewing Figure 8 on page 21, the improvement is from mid-lift on the opening through mid-lift on the closing. Figure 9 on page 21 shows again where pressure change affects the filling of the cylinder: by increasing mass flow.

PLENUM VOLUME

The plenum controls the pressure interaction (so-called "cross talk") or communication among the cylinders. A large plenum can decrease communication, while reducing the volume can increase the interaction. The geometry of the plenum can influence the reflection of the sound waves. Generally, more plenum volume will make more peak power but hurts throttle response and may negatively impact peak torque. To show the effect of plenum volume, the long runner case (computer model) with taper was modified to have significantly more plenum volume. Power at 7,600 increased by 0.9 bhp, again not representing a significant gain but verifying the concept. Figure 12 shows the pressure versus crank angle for this large plenum manifold. The pressure at the time of intake valve opening is less than in the other cases, due to the larger plenum volume.

The plenum acts as an accumulator and holds more mass at the time the valve is opened, allowing for a longer blow-down period than the other manifolds. (Note: This effect is also useful at lower engine speeds on restricted-intake engines.) Figure 13 shows mass flow versus crank angle for this case. Compared to the tapered and longer runner, the larger plenum delivers its benefit early in the valve event and just after peak mass flow. The mass flow versus crank angle graph shows the longer period of blow-down resulting from the increased plenum volume.

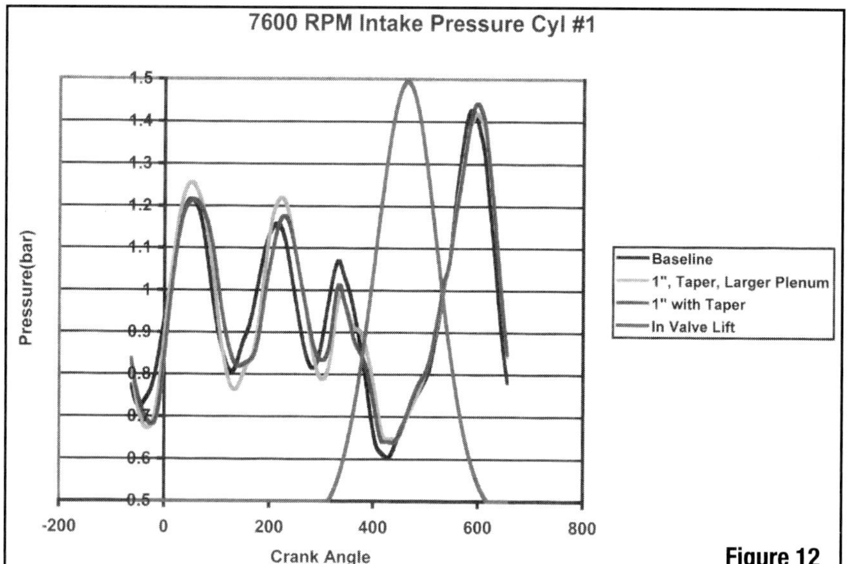

Figure 12

A FEW THOUGHTS ON CARBURETOR SPACERS

Think of them as tuning tools. Spacers can be configured to address specific conditions in a given engine package. Spacers can add plenum volume, providing benefits already listed in this book, and they are capable of helping diminish reversion pulses, improving combustion efficiency, and aiding stability of fuel delivery.

Sometimes spacers become diagnostic tools by pointing to other problems in an engine requiring companion parts changes, such as valve timing, spark timing, carburetor calibration, or manifold selection.

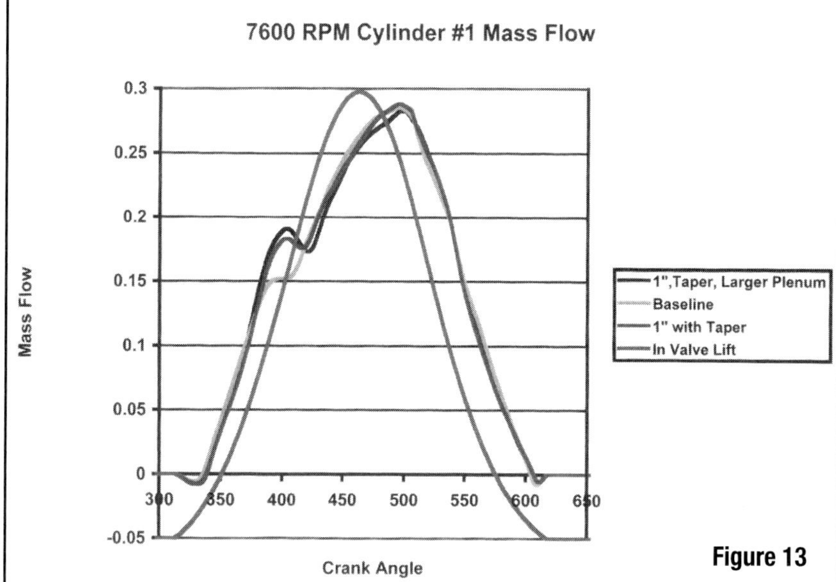

Figure 13

Chapter 4

BUILDING BLOCKS

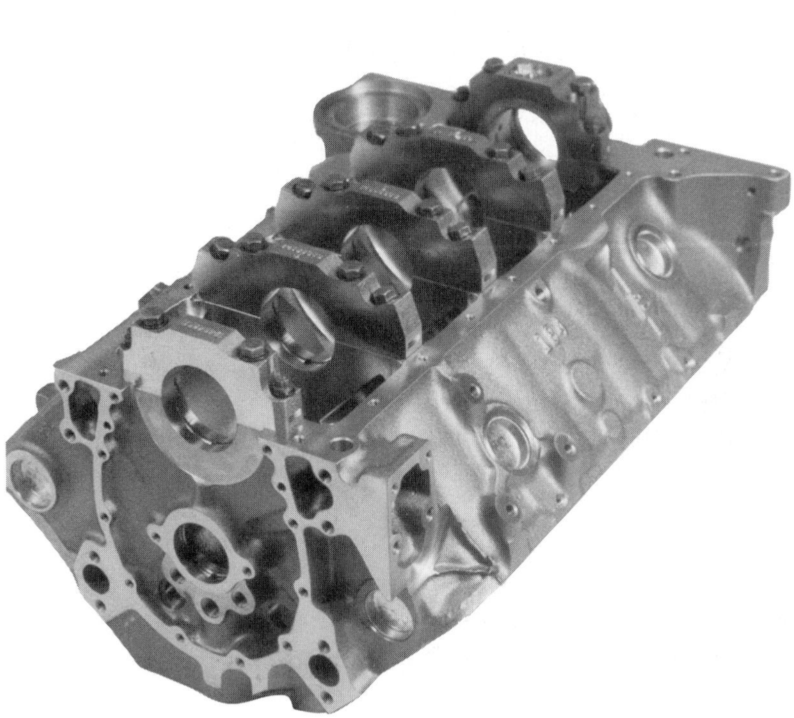

THE FOUNDATION OF A POWERFUL RACING ENGINE IS A WELL-PREPARED BLOCK

by Larry Jewett

The block is the foundation for an engine's performance. It is the vault of power and the encasement of durability. Proper care of the block is mandatory to keep the wallet from growing thinner.

The pressure builds. The drive to perform efficiently ramps up. Failure of one aspect could be devastating. It's a lot to ask of an engine.

You're asking the same from your engine builder, who's charged with creating the key to propel your racing career. After all, it's not what you have, but what you do with what you have. An experienced engine builder offers a competitive edge.

Chuck Jenckes works in the engine shop at Dale Earnhardt, Inc. (DEI). The team goal is to win races and bring home a championship. With their NASCAR Nextel Cup cars and countless racing activities on their plate, the shop crews have to be perfect the first time and better after that.

"The first thing everybody thinks," says Jenckes, "is that building racing engines—whether it's Nextel Cup or even Formula 1—is like a very dramatic, highly interesting business. The truth is, on an everyday basis, building racing engines is probably one of the most exacting, excruciating, if not almost boring, businesses. It's not excitement every minute. The actual process of assembling an engine is painstaking, and the level of detail you have to maintain is extremely high. It's rigorous, and it's not what people think. You're just out there every day making horsepower with everything you do. It all comes down to fundamentals.

"The interesting thing is that what makes a great Late Model Stock engine is probably a lot of the same things that make a great Nextel Cup engine—and that's attention to detail and making sure every detail is right to the extent that you have control. The difference is that in a Nextel Cup engine, you have a lot more control over what you can do, whereas in the Late Model Stock engine, because of finances and resources, you may not have that control."

To create a quality engine, you must have the best possible block. This requires preparation and a certain understanding of what is being accomplished. "While the block itself doesn't make horsepower in the sense that there are no moving parts in the block, all the parts move in the block," says Jenckes. "The block is like a house. If the foundation isn't good, you can't have a strong, good house. With a racing engine, if you don't have an excellent block, then you don't have the foundation to have the power to make it better."

BLOCK INSPECTION

The first step in block preparation involves machining. "Make sure everything is in the correct place and that things are square and true, like the crank centerline and the cam centerline," says Jenckes. "Make sure the crank bore and the main journal bore are square and that the line bore is properly done and placed."

Jenckes cites another area that's getting plenty of attention from today's engine builders: "There are attempts in all

A block should be visually inspected through each step of the process. Every step needs to be properly completed before any of the rotating assembly begins to make its way into its new home.

engines today to turn more shaft speed, so lifter plates are a huge issue. Are they placed correctly? Are they on the right centerline? Is the axis true? Check whether they are perpendicular to the cam or not and also the angle that they're indexed relative to that, and then the position of the lifter, left to right, front or back. All of those things are critical. If it's a roller motor, obviously the placement of the roller and how it's indexed against the camshaft—and if it's a flat tappet—is just as critical."

The crank centerline is a key element of the block preparation. "If you have an older block and you begin to bore it and set it up, you're actually moving the cam centerline and crank centerline closer together, but you compensate for that with different caps," says Jenckes. "Actually, the boring and honing process will get you straight and true, but the point is, even if you aren't re-indexing that bore, you need to check it. You need to make sure the journals are right. Don't assume anything. That's the key in almost all racing-engine preparation. Check every journal. Check everything. You have to check every bore of the mains. You need to check whether the crankshaft is square to the rest of the engine."

The block material plays a key role in the ability to develop a good base. The wall thickness of the cylinder liners will determine how much you can do. "Even if you're building a Street Stock engine," says Jenckes, "you want to have a good foundation. Take the block to someone with an ultrasonic checker. Most high-end machine shops will have them. It's worth the money to make sure there's enough meat in the block and that it's a sound block for the rest of your work. You don't want to spend all the money preparing it, ultimately finding it's thin, and when you go to bore it, you can't get as many rebuilds out of it as you would like. Obviously, as you go up in power, cylinder pressure, or torque, you have to worry about how thin the actual iron is."

CASTING

The manufacture of the engine block is not a perfect art. The best way to begin to understand the intricacies of a block is to understand how the block is created at the foundry.

Engine blocks are created when the material (in this case, iron) is poured into a mold. There are sand cores placed where the iron doesn't go. When the iron is poured, it shifts around the sand cores. "The water jacket is made up of a group of sand castings, pieces that are bonded together using some kind of agents like SO_2," says Jenckes. "The cores are placed in a cavity, and the metal is poured in a sand mold." Through this process, there exists a scenario known as "core shift." Core shift can create imperfection in the block. "Depending upon how the cores are located in the particular casting process, there's some slop in it," Jenckes says. "The whole process is done in a dark, nasty environment. Foundries are pretty much like the fiery furnaces of hell. The basic-production V-8 engines were made in the assembly-line process, so the cores may move around. With core shift, they're going to have cold spots, and the metal cannot be as uniformly thick. You might get thin spots in a block."

The problem may not exist at such magnitude for special-production blocks for pure racing engines, but the best builders will check and double-check each aspect of the block.

SIAMESE MYTH

Jenckes believes a siamesed block is as good as a non-siamesed unit if the block is prepared properly. "The old Chevy 400 (siamesed block) had a bad history. It had sensitivity for blowing head gaskets, and part of that was the production environment. If

There's a significant amount of handwork that needs to be done with an engine block. Many pounds can be cut away, but the fine adjustments that require detail are best done by a steady hand. Threaded holes need to be chased, and areas require deburring. A bare block can be a challenge that requires days of prep work.

you're using an old 400 block, the main thing is to check the block and make sure it doesn't have core shift. Check the deck surfaces to make sure they haven't been destroyed by being decked too many times."

WHERE TO START

In the process of learning more about block preparation, Jenckes gave us a view of the process used by shops such as DEI when a block arrives from the factory. A typical block is cleaned and inspected before it goes anywhere in the building. "Even though the company that does the machining for the GM castings is one of the best in the country, we still check the block," says Jenckes. "We check all the dimensions. Even after the inspection report comes in, we still check it. In a Nextel Cup operation, the block is one of the few pieces that's used over and over again. It's supposed to have a long life, and you want to know right away whether or not it's good.

"Whether you build a Nextel Cup engine or an engine for a Hobby Stock, it should be the same process. You need to clean the thing up and find out what you have. The fallout in a Nextel Cup shop is not very high. Usually, the amount of problems we would find would be minimal because they're expensive parts machined by the best people in the country. You'd expect there wouldn't be many problems.

"Find people who have been in racing awhile," Jenckes advises. "How do you find out anything in racing? Typically, you go to the people who have done really well and get their advice. You'll find that, in most cases, a lot of the work is done by a very few number of shops, and part of that is because racing work isn't as profitable as other types of work.

"The other thing is that you need to go and look at the shop yourself. Not every shop is going to be like a Nextel Cup shop where the floors are all clean and everything is painted white. You want to find a place that pays attention to detail. See if the bore gauges are clean and look like they're in good shape. All the equipment needs to look like it's in good shape. It may be old, because some of the best tools are old ones—and that's fine. The new stuff is great, but it's more the guys using the tool than it is whether the tool is brand-new or not. It says a lot if you go into a machine shop and you see all the tools are well maintained and the machines are all maintained."

DEBURRING AND LIGHTENING

"There's a big chunk of time spent deburring the thing," Jenckes says about the next step. "In a Nextel Cup shop, there are pounds of metal taken out of the lifter valley area. All over the block, all the sharp edges are broken off, and that kind of detailing is largely time-consuming. It's a very nasty job, but necessary. A lot of it is done with CNC, but the final deburring is not. It's polished—almost polished smooth—and that's not done with a CNC machine.

"There's some lightening work that's done. If you take out 10 to 15 pounds, you're doing it with a CNC machine. If you're going to take out a few

You can see evidence of the lightening that was done with this block. Metal was removed, which lessened the weight without compromising the integrity of the block. Make sure your block has the proper thickness after any lightening is done.

pounds, you're doing it by hand. After you get done with the lightening and the block is down to fighting weight, you're going to take all the edges off. You're going to polish the area where the oil is going to flow and the valleys. It's going to take a day or days, just in the handwork at the end of it. You want the oil to flow smoothly over the surfaces where possible, and you want to break all the sharp edges because the block is going to be handled for a long time. You don't want people cutting themselves. You check all the threads on all the holes. Every passage is brushed clean. After you get done with the grinding, you have to do the cleaning up. After it has been cleaned, it goes to lifter bore preparation, main preparation, and decking operations."

IMPORTANCE OF BEARINGS

Engine builders are hearing about the need for smaller bearings. Jenckes points out that main bearing sizes are dropping and rod pin sizes are getting smaller. "If that trend continues, ultimately what determines bearing diameter and width will be how much cylinder pressure you make and how fast you're spinning the engine," he says.

"As you reduce the radius, you reduce how fast you're moving, but you also reduce the load-carrying capabilities of the bearing. Because of better oils, better bearings, and better machining operations, you can get away with a smaller bearing today than you ever could in the past. There are frictional gains, and that's the primary reason for doing it.

"You have to be careful in doing that stuff. A Nextel Cup engine is built to run one time, and then it's rebuilt. If you have a racing engine that you're counting on running the entire season, a couple of horsepower gained through frictional loss probably isn't worth having to rebuild it five times as often.

"When you're building your engine, you have to be honest with yourself. What is it going to do? What is my goal? The block is the foundation for that, so you don't shortchange on durability. In the Nextel Cup world, the single driving factor is durability. One engine failure can ruin an entire season. In the 2000 season at Joe Gibbs Racing, there were no race-day engine failures for the No. 18 car, and that was part of the reason Bobby Labonte won the championship over Dale Earnhardt.

"Today, it is very important that you have no failures. Any Nextel Cup shop would sacrifice the opportunity to make power to gain reliability or reduce the odds of component failure. That's a routine decision that is made. People think that horsepower is king. It's not. It's durability. At home, when you're building an engine, you need to be honest with yourself. Running a smaller bearing could increase your chances of failure. That will just cost you a lot of money, and if you fall out of a race that you could have done well in, it's a very poor decision. A couple of horsepower is rarely going to make the difference in winning or losing a race. Finishing the race is always going to help you in whatever type of racing you do.

"The main thing you want to do is make sure the bearings fit properly. There are more important things than reducing bearing size."

The jury is still out on the need to Scotch-Brite bearings. There seems to be only one course of action that works best. "Follow the bearing

Modern cylinder boring machines can provide repeatability and attention to details.

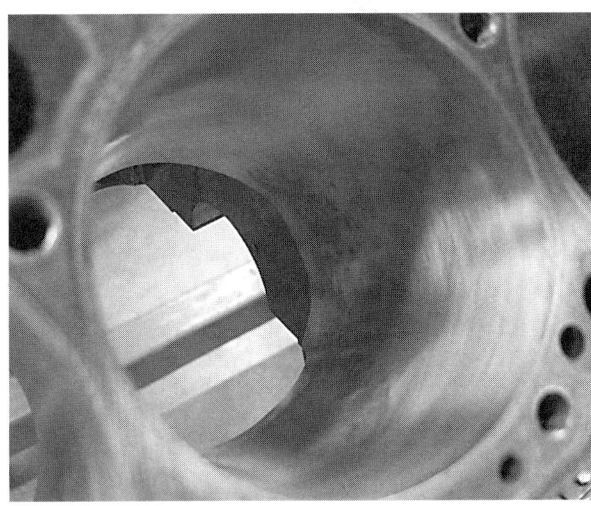

Your objective in working on the cylinder bores is to get them round and square. It's not a contradiction of terms, but an engineering feat that will yield the best results.

Racing-engine blocks can be composed of various materials. Popular choices include aluminum and iron. High-level engines are using a compacted graphite iron (CGI) for today's competition.

manufacturer's instructions to the T," Jenckes says. "If you want to have a bearing coated, that could be a viable thing. Coated bearings can buy you some added protection. It can be a life preserver in a bad situation, but you have to be very careful with that. You're not going to Scotch-Brite something that's coated, but cleaning them up is generally a good idea."

KNOW YOUR GOALS

Jenckes reminds engine builders to keep their objectives in mind. "Before you even get into block preparation," he says, "you have to ask yourself two questions: How much money do I have to spend, and how much maintenance can I do? If this engine has to go together and stay together for an entire season, that's one consideration. If you take it apart a lot more and service it, then you can do more things to make power. It's not just the cost of the power; it's also the cost of the maintenance. What happens in a lot of situations is that guys just don't have the time to maintain them. They don't have enough engines to be able to rotate them, and that's where the added cost comes in, not just chasing horsepower. The most important thing to be successful in racing-engine development and building racing engines, for either a customer or yourself, is to be honest and realize how much maintenance you have time for. In a lot of cases—again, just like in Nextel Cup racing—it may be more important to spend the time preparing the chassis and preparing other things, or getting more seat time if you're a young driver, than to spend time on engines. In Nextel Cup racing, engines are not the most important thing. They're important. It's still motorsports, so the motors are still important, but it's not the most important thing. You need to prioritize your resources and decide early in the game what you are building. If you've got the money to build for power, but you can't take the thing apart and service it properly, you're better off taking a more conservative approach."

THE NEXT STEP

We have lightened the block, cleaned it up, and done a thorough inspection. Where do we go next?

"Basically, after you have the fundamentals done, you can talk about the bore itself," Jenckes says. "There's a lot to be said about bores. The most important design element is, at operating temperature, to have a round and square bore or have a bore that is as close to cylindrical as possible. The piston runs up and down, and if the bore is round and straight, you can run tighter piston-to-wall clearances. Your rings are going to perform better. You're going to have less friction. This whole interaction between the block and the piston and the rings has a lot of performance that can be left on the table. The entire combustion process has to be sealed by the rings. If the rings aren't working properly, you're going to be losing power. Having good sealing is critically important to making the engine work."

Jenckes points out that great strides have been made to make bores cylindrical. "You find pretty quickly that these hot honing techniques people are talking about, even used at the Street-Stock level, are going to get you power. You need to hone the block at operating temperatures, and you need to have the honing oil hot, not just putting water through the water jacket. The oil needs to be heated to anywhere between 150 and 190 degrees. Getting the block up over 120 is probably a good area to be operating, but if you can get it to 150, it's better. You have to have hot honing oil. If you heat the block up and douse cold honing oil on it and then hone, localized cooling is going to distort the bore. With a production block, you have all these residual stresses built up in the block, and the bore distorts tremendously. We're talking about multiple thousandths of an inch, like five thousandths. When you have piston-to-wall clearance of eight thousandths and a distortion of five, that's a huge percentage."

The goal is the perfect circle—actually, a perfect cylinder. While being round and square may sound like a contradiction, it's the ideal objective. "The cross section is a perfect circle," Jenckes states, "and it's perfectly square because the cross sections are true, so it's round and square. It's a perfect cylinder."

Boring for cam bearings is a precision step in which modern engine builders rely on high-cost machinery. When building an engine, there's plenty of emphasis on the operator as well as the machine.

MORE ON HOT HONING

Jenckes believes that hot honing, when done properly, can be beneficial. "It is certainly a technique that has been proven fundamentally sound at all levels," he adds. "Is there one brand of hot hone that's better? No. It's not to say a guy at a machine shop can't build his own hot honing apparatus. There are many on the market, and people need to evaluate them. It is absolutely a technique that is good at all levels because it provides the foundation for what is going to come later, which is how the rings are going to perform and how the pistons are going to work. The trend is toward lighter pistons and higher shaft speed."

MATERIALS

Today's engine block can be made from one of a number of possible materials. While many of the racers are dealing with simple gray iron, the composition of the block must be taken into consideration.

"There's something called compacted graphite iron, or CGI, that has about 50 percent more tensile strength than gray iron," Jenckes points out. "That material is being used in all levels of motorsports for high-end racing blocks. That's a good foundation.

"There are different types of iron—gray iron, nodular iron. All of these irons have different chemical make-ups and different properties. Depending upon where the block was made, the foundry, and the metal used, you can have different performance from them."

Each type, from the CGI down to the gray iron, has its unique properties. "Most of the blocks ordered in Cup today are CGI, and it's harder to machine," says Jenckes. "The big downside for compacted graphite is that you have to tap it. You have to be careful because you'll be breaking taps. It's tough to machine, and it's also much more sensitive when you hone it. The honing aspect of it becomes critical. In most Street Stock stuff and Late Model stuff, you're limited in what block you can run."

BIGGER PACKAGE

As the block preparation work continues, the elements to produce power start to come together. As Jenckes emphasizes, good ring seal is critical.

"You have to look at the cylinder kit, which is the bore, piston, rings—those components," he says. "Some of the things that are critical in getting good ring seal are the finish of the bore, what level you hone it, how you hone it, even what you hone it with, and not the least of which is the guy honing it. Machines make a lot of it easier. I can remember a guy standing on top of the block with a 1/2-inch drill motor, and that's how blocks were honed. It worked fine. Part of that was the guy operating it. He was so skilled he could feel a block burnishing just by the increase in the motor speed. When they came out with machines, you could watch a load meter run while you were honing the block. There are a lot of machines out there now that can help the average guy hone a block than ever before."

COATING SLEEVES

According to Jenckes, there has been some experimentation in applying coatings to cylinder sleeves.

"Nickel silicone carbide is a ceramic that coats the parent metal of the block, or the sleeves are coated and placed in the block," Jenckes explains. "Coatings are difficult. Whenever you coat something, you have to be very careful that the coating sticks. There's a substrate on the parent metal you're applying a coating to that's an integral part of the chemistry. In a cylinder block, the iron itself is not homogeneous. As they pour the iron and it cures, it's not consistent throughout the block. The chemistry of applying a coating to an iron block is problematic at best. A sleeve is easier to coat because the material can be completely homogeneous. It can be a tube that is specially made to be coated. The chemistry is much easier, but even then you can have coating slips, and that can be a problem.

The four-bolt main is a preferable way of making long-lasting power. You can get power from a two-bolt, as proven in the drag racing world, but endurance can become an issue.

"It used to be that people didn't use sleeves, thinking a sleeved block was a bad block. Today, they're different sleeves. Some of the sleeves they're running in motorsports are 0.0004 inch, and that's not much of a sleeve.

"There are some significant advantages to running sleeves. Many people don't believe in them and say they're a bad addition, [but] many people have had good luck with them. So we're not going to find anyone saying sleeves are perfect, or they're great, or they're junk. There's still a lot of debate over it. Like anything else, you can build winning engines with or without them, but you have to be careful how you handle each side of it.

"As far as friction reduction by using a Nikasil coating, I don't know that I have personally seen power gains just by doing the Nikasil. In and of itself, it may not be a power gain, but you may be able to build power you couldn't have without it by using other technologies."

SEEKING THE RIGHT BORE

Jenckes found a possible advantage of using sleeves: They come with bigger bores. "Today, people find that particular bore and stroke combinations are very effective," he adds. "Typically, people are running bigger bores and they find that, no matter what the application, a bigger bore may be helpful. I'd certainly say that's the trend in Nextel Cup. Obviously, if you want to make a block last a long time, you can't go to the big bore size immediately. You'll get one rebuild, and then it's junk. A sleeve can allow you to go to your optimal bore size. When you have problems, just replace the sleeves.

"Some teams in Nextel Cup use sleeves, and some don't. In Nextel Cup, the well is deep money wise, but it's not endless. Consider that you can use a smaller number of pistons and crankshafts, and that means less money tied up in crank and piston inventory. Some teams might approach it from that standpoint."

POWER FROM BORE CENTER

Bigger bores have proven to have better results in most cases. "A larger centerline is preferable because it allows you to have a bigger bore," says Jenckes. "A Ford block has an inherent handicap compared to a Chevrolet or the current Dodge because of the smaller centerline. Having more metal there, other than a weight handicap, means it's more stable and it can allow you to run a bigger bore.

"Most racing has some kind of displacement rule. As long as there's not a bore size rule, you're probably better off running a bigger bore and shorter stroke if engine speed is not limited. If you are limited in engine speed for some reason, say an rpm limit, then always going with a bigger bore may not be the way to go, depending on how low the speed is. At higher shaft speeds, bigger bore engines help the cylinder heads work better. In Nextel Cup, 4.185 inches is the biggest bore size you can have. Even if you have 5 inches between bore centers, it doesn't help you other than the fact you've got more metal there to be stable. Then you have to worry about the weight. The trend has been toward bigger bore cylinders. GM is working on a new, next-generation Nextel Cup engine with a bigger bore center."

CAM AND CRANK PLACEMENT

There's a design movement to position the camshaft and the crankshaft higher in the block. This begins our look at valvetrain geometry.

"You move the camshaft up to shorten the pushrods," Jenckes explains. "Obviously, with very high spring pressures, a shorter pushrod is a stiffer pushrod and a lighter pushrod in general. In the valvetrain, you want things stiff and light. There's some kind of optimal stiffness-to-mass ratio that you're going to go after. Having a shorter pushrod that is directionally correct may actually be negatively correct for the valvetrain geometry. You have a shorter distance to go, so you make more angularity in the pushrod. Ultimately, time has proven that the shorter pushrod path is a better path to go down. In very few cases would you find arguments where you want larger pushrods.

"The second thing you'll find is that the camshaft diameter goes up. It increases the radius of the cam on the base circle, and the raised curvature goes up, so it is directionally correct for valvetrain dynamics. When you have a larger base circle, you have a lot more room to work with the cam profile. For the transitions, a larger radius can be smoother than a very sharp one. Having a larger base circle means you can have better valvetrain performance."

MAIN STRENGTH

A final topic of consideration is the number of bolts for the main caps in the block. It's the classic two-bolt versus four-bolt argument.

"A lot of power can be made out of a two-bolt main," Jenckes offers. "If you work in the world of drag racing—stock eliminator stuff—you see this. Drag racing forces people to use production blocks. With the two-bolt main stuff, they make ungodly horsepower. The question is, how long do you make it? In a Nextel Cup engine, in round numbers, you're making 800 hp, and there's a lot of cylinder pressure. Because of that, you have to retain it over time. And that's where you get into the two-bolt versus four-bolt.

"A four-bolt is a stronger combination if done properly. Almost all Nextel Cup engines use splayed main caps, which are much more satisfactory than the vertical four-bolt. Again, you're tying in more material. It's all about stability. It's more about peak performance in the long run than it is about the capability of a two-bolt block that can make power for a short period of time. Like I said, the stock eliminator blocks are a perfect example of two bolts that make power—some make as much as a Nextel Cup engine. They just don't make it last as long."

Chapter 5

MIX 'N' MATCH

How to Combine Engine Components for Maximum Horsepower

Text & Illustrations by Jim McFarland

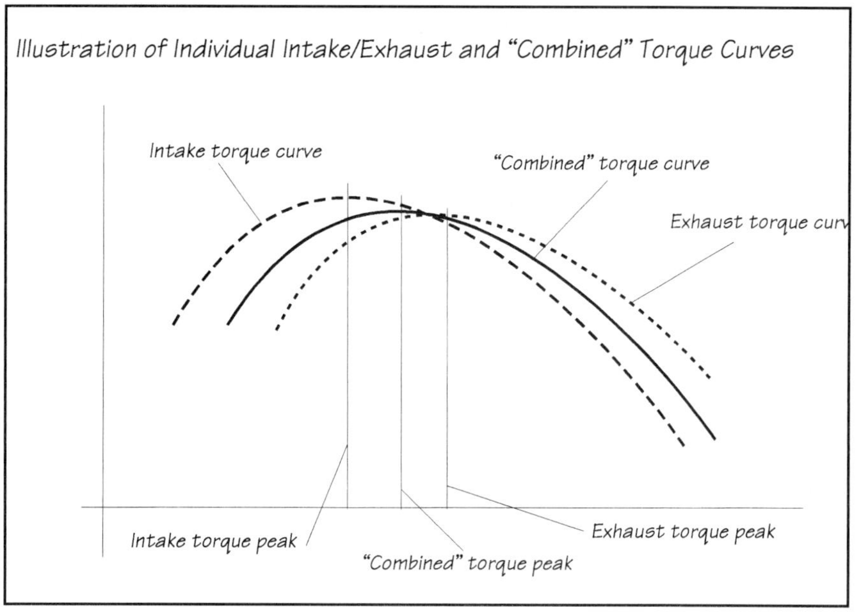

These days, engine-building is more an exercise of assembling ready-made parts into a working unit. The trouble is, very rarely is any one part the best choice for every application.

VOLUMETRIC EFFICIENCY

To understand how to mix and match engine components to get the best final product for your application, it is critical to understand the meaning of "volumetric efficiency" and how it relates to torque (or power).

Technically, volumetric efficiency (VE) may be defined as the ratio of actual air capacity to ideal air capacity. Perhaps more simply, it is a measure of the volume of air in an engine's cylinders at any given rpm compared to the physical volume of the cylinders (total piston displacement).

For example, if 80 percent of actual piston displacement (volume) is filled at 5,000 rpm, we can say the engine is 80 percent volumetrically efficient at this engine speed. In theory, 100 percent VE represents the maximum obtainable under the influence of atmospheric pressure. However, it's possible to achieve more than 100 percent if certain conditions in the intake path are met. Induction system "tuning" can provide this benefit, based on proper dimensioning of the inlet path.

Why is volumetric efficiency important? Clearly, the more air an engine can ingest throughout its rpm range, the greater the potential for optimizing torque and horsepower. The key point in volumetric efficiency affects and is affected by the performance of several major engine components. Because of this, their selection and functional integration is important to finding winning power.

We also should point out an important difference between torque and horsepower: Torque is considered an independent variable—the result of volumetric and combustion efficiencies. It is the potential to produce a twisting force. Torque and volumetric efficiency are closely related. Horsepower, since it is calculated using values of torque, is a dependent variable and is the result of torque acting over a period of time. Because of the close relationship between torque and volumetric efficiency, it is vital that engines be built to optimize VE in the range of engine speed where they are operated most of the time. If you are racing at a short track and always grunting out of the turns, you need your torque on the low end. Big tracks where the revs are high need the torque on the top.

There are several engine components and functions that affect VE. In particular, the dimensions of intake- and exhaust-flow passages (especially intake manifolds and exhaust headers), valve timing, mechanical compression ratio, rod-to-stroke ratios, the carburetor, and air-inlet system sizes are important to consider during parts selection. What you need to know before you start is where the engine needs to be optimal VE, maximizing torque and performance. If conditions require an engine to produce torque over a wide span of rpm, certain compromises are necessary that aren't if you are building for conditions that allow an extremely narrow range of rpm. This will become more apparent as we discuss individual engine components and systems.

If a broad torque range is required, it's possible to tune the intake and exhaust systems separately from each other. Simply stated, the intake system can be tuned to a higher (or

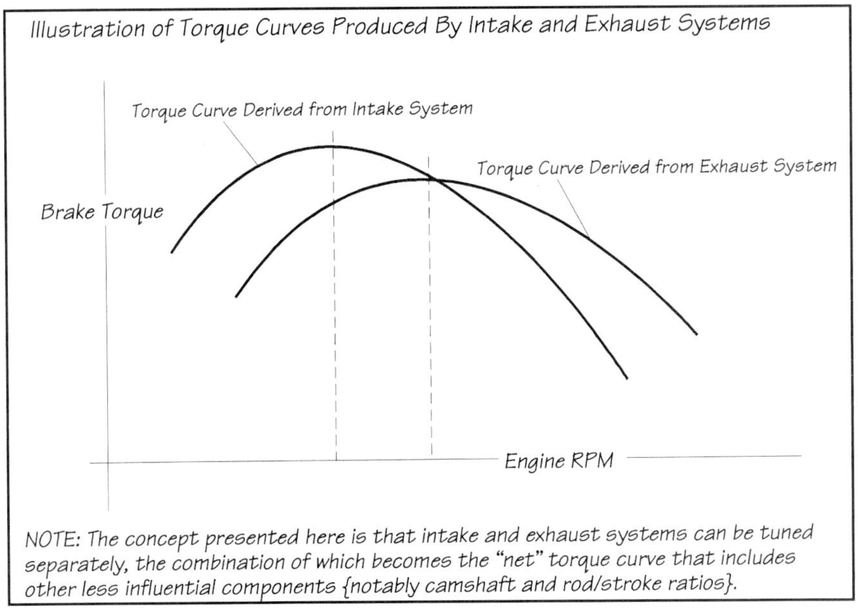

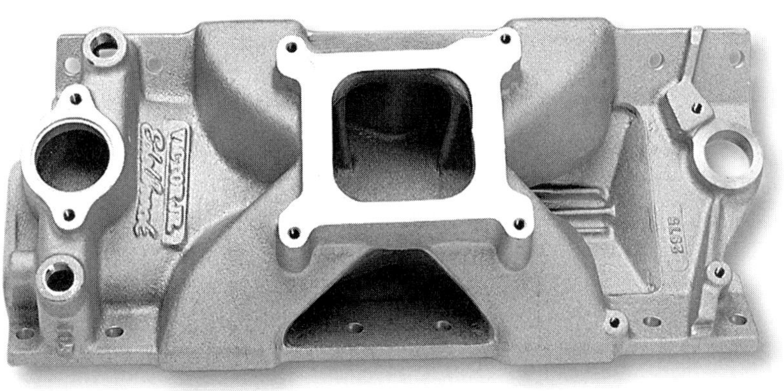

The Victor Jr. by Edelbrock is an rpm-flexible intake manifold directed to a wide variety of circle-track applications. Its ability to enhance torque over an extended range of rpm enhances various on-track situations where good torque is necessary. Depending upon the size of carburetor used and the range of rpm, this manifold responds well to carburetor spacers; open and of two inches height.

lower) rpm range than the exhaust. This tends to soften, or broaden, the span of engine speed where torque is available. As the rpm band becomes narrower, intake and exhaust systems should be tuned to more nearly the same rpm range.

THE ROLE OF RPM WHEN INTEGRATING PARTS

Perhaps the single most important element when choosing and matching major engine components is careful identification of the rpm range in which they will need to perform. Engines with "mis-matched feet" seldom produce winning power. Also woven into this equation is the adage that "bigger isn't always better." That said, let's examine a few rpm-related issues.

Airflow vs. RPM

Since VE is key to making torque and is related to airflow, we need to discuss airflow vs. rpm. Certainly, as engine speed increases, so does airflow rate and volume. The pumping motion of pistons creates in-cylinder pressure conditions lower than atmospheric pressure, causing the outside air to rush in to fill the negative space in the cylinders and produce an inlet flow. As the rpm range in which an engine will be run widens, airflow velocity and volume become increasingly critical to good volumetric efficiency. Therefore, the sizing of intake and exhaust paths comes into play.

At any given rpm, the cross-section area of a passage, not its length, affects flow rate. Remember, if an engine is to be operated primarily in the 4,500–7,000 rpm range, it is unimportant how much VE it produces at 7,500. Intake and exhaust passages should be sized according to the applicable span of rpm. The same holds true for other flow-related components, including camshafts, cylinder heads, and carburetors.

MAJOR PARTS THAT AFFECT VOLUMETRIC EFFICIENCY

While each of the following components may be selected on the basis of individual performance capability, it is more important to match them in terms of mutual compatibility—particularly with regard to matching rpm ranges. Keep in mind it is possible for an intake or exhaust system to provide separate boosts to an engine's overall torque curve. Therefore, each system can be selected to address different rpm ranges.

Intake Manifold—Take a moment to consider

As engine displacement or maximum rpm increases, the slightly larger (compared to the Victor Jr.) Edelbrock Super Victor provides torque increases in the mid- and upper-rpm range. As with the Victor Jr., the combination of short and longer runners and their different cross sections tends to broaden and flatten the overall torque.

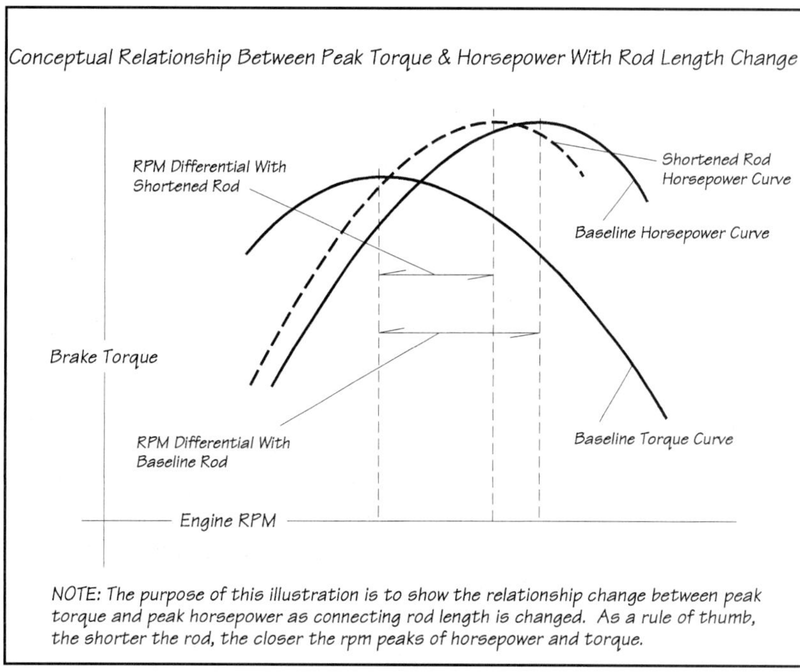

the following set of conditions: Visualize an engine of 355 ci, running 4,500 rpm at peak torque (for the intake manifold). The manifold has runners with 2.75 square inches of cross section. The flow velocity in the runners (at peak torque) is 240–260 feet/second. (This number is generally the desired rate of peak torque for any displacement.)

Now, assume we switch to a manifold with 3.0 square inches of runner cross-section and no other changes. In order to reestablish the 240-260 feet/second flow rate, rpm must be increased. The loss in flow rate at 4,500 rpm indicates the engine is now more volumetrically efficient at a higher rpm, moving the peak torque to a higher engine speed. Had we needed to improve VE at 4,500 in the first place, installation of the larger manifold would have likely diminished on-track performance in this range.

Note that we have not discussed intake runner length. Changes in passage length have little to do with peak torque rpm. Rather, lengthening or shortening the path tends to pivot the torque curve about the peak, neither raising nor lowering the rpm point at which the peak occurs. As a result, long runner manifolds with properly selected cross sections tend to produce good off-the-corner and passing torque, typically over an extended range of rpm. Shorter, larger runners work well in the higher rpm where the on-track changes are smaller. Intake manifolds with multiple runner lengths, cross sections or both tend to reduce peak torque values while broadening the overall curve.

Exhaust Manifold (Header)—There are certain similarities in the design and function between intake and exhaust manifolds that are helpful when selecting and matching the two. In particular, the primary pipe diameter of the header is linked to peak torque rpm similar to intake manifolds. The following simplified equation for determining the relationship between pipe size, cylinder displacement, and the peak-torque rpm point is helpful for determining your needs:

Peak torque rpm = (Pipe cross-section area x 88,200) ÷ cylinder volume

If you transpose this equation to determine the primary pipe size you need and and know where you want your torque peak, the format becomes the following:

Pipe cross-section area = (Peak torque rpm x cylinder volume) ÷ 88,200

In the case of the first form of this equation, it is possible to determine at what rpm the exhaust system will contribute to a torque boost. Based on engine size and a set of headers you may already

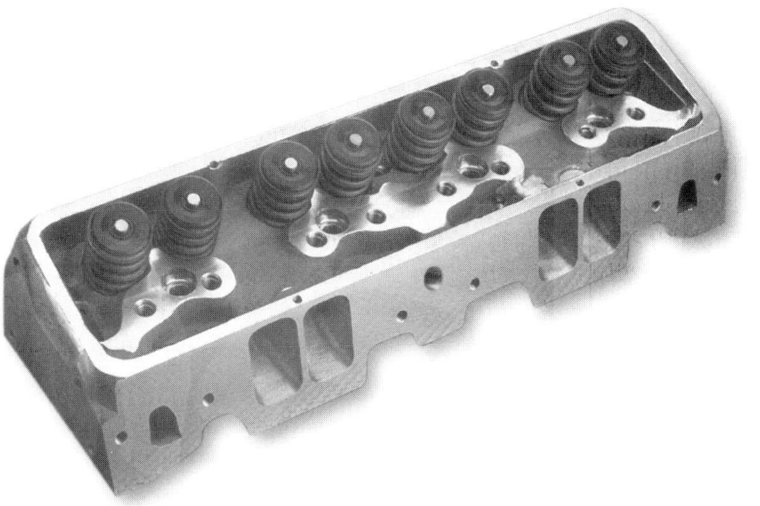

Currently, the Air Flow Research approach to cylinder head modification focuses on high-velocity port flow without minimizing total flow volume. Specific improvements to torque output and throttle response is a result of this concept in port design.

want to use, this computation will show where in the rpm range the exhaust system is most effective.

The second form of the equation suggests that you know the engine size and rpm at which you want an exhaust-system torque boost. It tells you the cross section of the header required to achieve that peak-torque rpm point.

By way of comparison, intake-manifold selection (or analysis) can be made in a similar fashion. A fundamental difference is that most intake manifolds have runners that include taper, unlike header primary pipes. In the case of intake manifolds, reasonably accurate information can be computed by averaging runner cross-section areas. To do this, measure the inlet area and outlet areas, then take a mathematical average. Use this information in the previously indicated equations for evaluating intake-manifold performance relative to peak-torque rpm. It's not deadly accurate, but it's clearly in the ballpark.

Concluding our discussion of header evaluation/selection, let's take a look at header-collector function. Collectors provide a torque boost below the peak-torque rpm point. Since the majority of oval-track engines are required to produce torque below the peak, collectors can be beneficial. On the other hand, if track/gearing/rpm conditions cause an engine to produce excessive torque in this range, shortening or removing collectors can be helpful.

From one perspective, the addition of a muffler tends to create a collector function. The less backpressure a muffler offers, the more it will behave as a collector. Keep this in mind if you're required to run some form of muffling device.

Finally, before leaving this segment on headers, let's discuss how this little math model can be used to determine if an engine has been underported or overported, with respect to the sizing of intake and exhaust systems. You'll need some engine-dyno data to make this work.

By measuring intake-manifold runner and header primary-pipe passage cross sections, we can determine the two rpm points where the intake and exhaust torque peaks should be developed. Let's say the intake is calculated at 4,300 and the exhaust at 4,600. By averaging the two, which is what the engine tends to do in producing a single torque peak, we determine 4,450 to be the point of net peak torque.

If the engine-dyno data shows a peak higher than this, the engine is likely overported, suggesting there may be too much valve timing or port sizing for where the math model indicated the peak torque should be. Should the data show a lower peak-torque rpm, we can assume the engine is underported, requiring an increase in volumetric efficiency at the rpm calculated by the model.

Also, while examining engine dyno information, there's another rule of thumb that helps in determining efficiency of your engine parts packages. The lower the brake-specific fuel consumption (bsfc), without an attending loss in power, the higher the combustion efficiency, or ability of the engine to convert air and fuel into heat. So, as engine tuning efforts are made during dyno evaluation, working to minimize bsfc while optimizing power is a generally good approach. You may be interested to know that the on-track performance of two engines with relatively equal power levels will usually be better in the one with the lower bsfc values.

Camshafts—Armed with information about where in the rpm range you intend to run the engine, most camshaft manufacturers can provide ample technical assistance. In some cases, available PC-based software will enable you to improve the

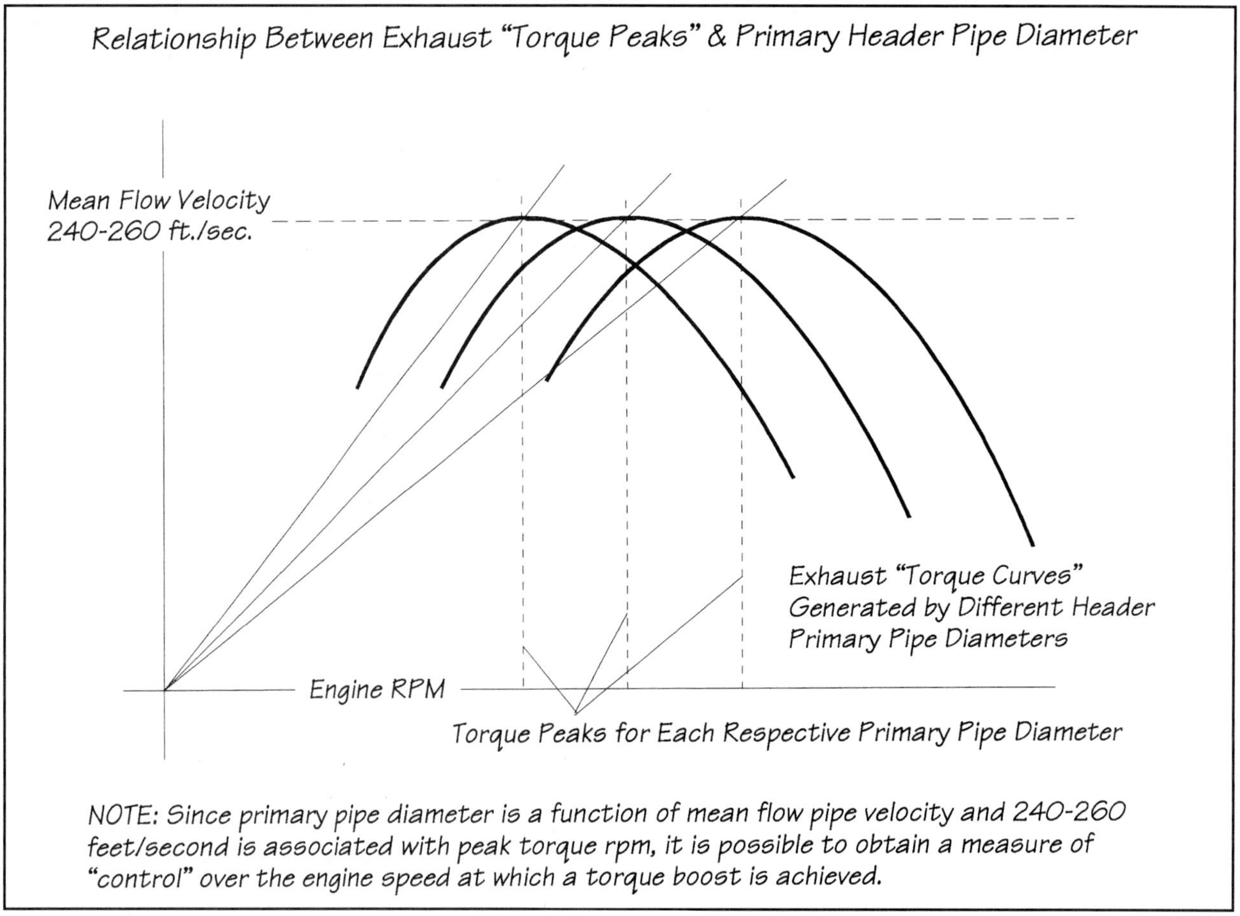

precision of a given camshaft selection. In fact, the use of software such as Desktop Dyno, Engine Expert, and Dynomation can improve the accuracy of your camshaft selection. Each of these programs includes some form of valve-timing analysis that can sharpen the selection process.

Cylinder Heads—Based primarily on port sizing, cylinder heads can be selected to address certain ranges of engine speed. In much the same fashion as intake manifolds or headers, port sizing, and net flow can be linked to the rpm range in which they are the most efficient. Even though the lengths of these passages, compared to intake-manifold runners and header pipes, are relatively short, cross-section area is still a factor affecting peak-torque rpm points.

However, keep in mind that because both intake and exhaust ports involve substantial changes in flow direction (into and out of the combustion space), it is typically necessary to produce shapes that depart from uniform section area. Considering the comparative-length relationships between ports and their extensions (manifold runners and header pipes), providing efficient transitional flow in and out of the cylinders is more critical than port tuning.

The selection of cylinder heads is sometimes confusing with respect to port volume comparisons. For example, consider the many possible ways an 180cc intake port can be configured. Many different ports can be shaped that measure 180 cc, and every one of them may flow differently. So, while it's possible to make general comparisons of ports by the use of volume dimensions, take care to find out what you can about port-flow capability. As a rule, the higher the port flow, the larger the port—and the higher the rpm range where it produces good VE. That's not an ironclad rule, but it's a pretty good guideline when shopping for and integrating cylinder heads into a parts package.

Carburetors (unrestricted engines)—Again, you can relate carburetor sizing to net volumetric efficiency and where in the rpm range torque is required. There are some mathematical formulas you can use here, too.

Because an engine will displace half its total piston displacement with each crankshaft revolution, first divide the cid by 2. Then, because we're dealing with cubic feet units (cfm) with carburetor flow, divide half the cid number by 1,728 (the number of cubic inches in one cubic foot). Multiply the answer by the maximum rpm at which power is required. Then multiply this number by the VE (assumed) at the chosen rpm. The answer will be the carburetor cfm required.

Here are some sample numbers. If the engine is 350 cid, half this displacement is 175 ci. Dividing this number by 1,728 yields 0.101 cubic feet per revolution. If we'd like adequate airflow to be 7,000 rpm, 0.101 x 7,000 obtains 707 cfm. If we assume the engine to be 90 percent volumetrically efficient at 7,000 rpm, the required carburetor size becomes 707 cfm x 0.90, or 636 cfm.

Now, let's consider why an engine might produce more power with a larger carburetor not flowing at its full capacity. Engines often become "overcarbureted" but still show an increase in power over what a mathematically sized carburetor produces. The most common reason for this is rather simple. The air/fuel mixtures are required to make a rather severe directional change between the carburetor exit and the entries to the intake manifold runners.

This transition can be aided by slowing down the mixture as it makes the directional change, which helps reduce the mechanical separation of air and fuel. The carburetor exit-flow speed (all else being equal) is reduced as the throat size is increased. So, it is possible to improve air/fuel mixture quality in this transition simply by using a carburetor with larger throats. Reducing the impingement of wet flow against an intake manifold's plenum floor is the issue. This, by the way, also can be addressed by raising a carburetor with spacers—rules and engine-compartment space permitting.

Mufflers—We're including these parts simply because there are times when mufflers affect net volumetric efficiency in the form of backpressure.

As backpressure rises, combustion efficiency (due largely to contamination by unscavenged exhaust gas) tends to decline. Stated another way, the effects of backpressure can be limiting to effective and efficient rpm development. Free-flowing mufflers also allow the use of camshafts with shorter overlap periods, closer to what you'd expect in the absence of a muffler.

So, while in the process of putting together your engine package, include the possible effects of mufflers that cause backpressure at high rpm. Not all mufflers do this, so consider the effects of many designs. If little or no significant backpressure can be expected, concern for negative effects can be minimized. If rules require mufflers, make your choice wisely. There's no reason to build a 7,500 rpm race engine if exhaust backpressure hurts power at 6,500. If this situation is unavoidable by the choice of a backpressure-producing muffler, simply assemble a group of parts that works below the limiting rpm point.

SOME CONCLUDING THOUGHTS

Fundamental to the selection of parts that are functionally integrated is making certain they all are designed to work within the intended rpm range. As simple as that sounds, it's a rule of thumb often overlooked. Decide on the range of rpm in which the engine will be run. Include both low- and high-rpm points that relate to off-the-corner torque (low rpm), passing power (mid rpm), and a maximum rpm that will allow deep-corner performance to further reduce lap times.

Many parts manufacturers base their parts recommendations on an anticipated range of engine speed. Use this to your advantage. Taking the time to nail down a realistic span of rpm can be a significant step toward proper parts matching. It isn't a bad idea to be conservative here. Engine-building with a broad torque range also can fit many types of track requirements because torque will accelerate a car, while horsepower can make it fast. Good end-of-straightaway speeds won't get the car around the track as quickly as gobs of torque. Selecting parts that accomplish the torque issue is key to race-winning performance.

Chapter 6

BLOCK HONING

MAKING CYLINDER BLOCK BORES CONCENTRIC AT OPERATING TEMPERATURE

Text by Jim McFarland
Photography by Dennis Wells

It is widely recognized that cast, machined metal parts tend to have "memory" created from internal stresses. In the presence of temperature changes, these stresses often cause dimensional and structural shifts. Racing-engine builders know about this condition, which can have a direct bearing on cylinder bore concentricity. In fact, we often spend extra time and funds to produce round cylinders in an attempt to produce bore dimensions equal to those that exist after an engine reaches operating temperature and cylinder pressures.

More specifically, cast and machined bores in a cylinder block are subjected to thermal and load distribution that add complexity to the goal of maintaining concentric bores in a running and highly stressed engine. Like it or not, there's some "Kentucky windage" built into the task of producing dimensionally correct cylinder bores that model those subjected to in-use heat.

We know the distribution of operating temperatures and combustion pressure is anything but uniform in the metal structure surrounding cylinders; three-dimensional growth of block material occurring in this proximity is unavoidable. So we've contrived ways to preload blocks with honing plates by heating the material or by using other methods intended to anticipate how a given block will experience bore distortion once it is assembled and running . . . sometimes on the assumption that all bores will distort the same, which they do not.

None of this is to suggest such methods are not effective, because they have been and are. Rather, these comments go directly to the problem of devising improved ways to compensate for the "effective" distortion that is inescapable under the influence of operating conditions.

Enter Stephen Kuzara of Sheridan, Wyoming-based Millennium Products Corporation. He has engineering training as an engine builder/machinist and is running his own machine shop. That's not an unworthy combination.

In prior conversations he'd had with Smokey Yunick (who had tried the process of "hot honing" during the '50s), Kuzara explored ways to circulate heated liquid through the cooling jacket of a cylinder block during finishing honing operations.

"Curiosity and frustration arose from chasing the final few tenths in the honing process," Kuzara says. "During the early '70s, I worked with a machinist friend who owned a CK-10. I began wondering how important those last few tenths were in the overall effort of achieving precision-machined blocks. I'd already observed how the stones would heat the bore surfaces and create dimensional changes, so it seemed logical some major changes would probably occur at engine operating temperatures."

In fact, Smokey had abandoned the process (although in the '70s he'd shared with this author the benefits he'd projected would occur), considering the way he'd tried the method was

FROM DENNIS WELLS' PERSPECTIVE

Wells is a no-nonsense builder who changes his techniques only when the benefits are worthwhile. Following are his observations on the process, after having used it on each of the cylinder blocks presented in the accompanying data charts.

Why would you substitute this honing process for any you've previously used?

"Simple. In order to have straight and round bores when the engine is operating at running temperature."

In your experience thus far, what dimensional advantages have you discovered this process provides compared to methods you are currently using?

"Most cylinders grow an average of 0.0035 inch when a block is heated to 200 degrees F from 65 degrees F, but they don't grow in a concentric manner, and this is an important factor. Therefore, the bores are basically different sizes from top to bottom."

Until being introduced to hot honing, what steps did you follow in cylinder-bore preparation?

"Previously, we installed a torque plate and did a four-step honing process."

From what you've seen, what improvement in power can you estimate compared to your previous practices?

"There should be between a 1 and 2 percent power increase, and you should need to change rings less frequently to obtain proper ring seal."

What problems or other benefits do you anticipate engine builders will experience while adapting to this process?

"Well, here's what I've learned so far: An exhaust fan will be needed to remove the smoke from the heated honing oil. Water and a drain hose will be needed at the machine. The water is very easy to remove from the block without getting into the honing machine. All you need to do is remove the quick-connect coupling from the fill side, connect the drain hose at the bottom of the tank, and blow air into the fill hose.

You need to measure cylinders when they are cold, heat the block to operating temperature, and recheck the bores. Subtract cold numbers from hot numbers; this is the amount you add to the dial bore gauge for the finished size. If this method isn't used, the pistons will stick in the bores when cold. It's advantageous to heat the block, thereby identifying any leaks. I've found leaking freeze plugs just by heating the block.

After you wash the block, there'll be less black residue to wipe away, as there is with cold-honed blocks. If you find a cylinder that grows an unusual amount, I would not use the block. This can save money from problems that appear later.

Finally, I discovered that painting a black spot on the side of the block helped radiate heat and trigger my infrared heat gun, making it easier to know the block and cylinder temperature during the honing process."

cumbersome, "messy," and "took too long."

Kuzara appears to have been more persistent. "My first setup was hooked up to our water heater, with one hose running water into the block and another running out to the drain," he says. "The outlet hose had a pressure gauge and a valve to help monitor the pressure and control the flow. Water temperature during the first test I ran was at 180 degrees F. The distortion measured was significant enough that I quickly concluded the last few tenths I had been chasing 'cold' were lost in the greater distortion I was seeing 'hot,' which was several times that produced by torque plates."

Interpretation? While torque plates do preload and create bore distortion, thereby providing data that enables improved honing precision, the process of hot honing (using torque plates) appears to create and allow material movement that closely relates to cylinder dimensions of a running engine at temperature.

TYPICAL BORE DIMENSIONAL VARIATIONS

For purposes of this book, four small-block Chevrolet cylinder blocks were selected from inventory at Wells Racing Engines in Duncanville, Texas: a stock 350, a stock Bow Tie, a 0.060-overbored Dart, and a Dart Little M (lightened). On the next page, there are charts of data provided, each of which presents cylinder bore dimensions obtained "cold" (65 degrees F) with a torque plate and "hot" (200 degrees F jacket temperature and 195 degrees F torque plate coolant temperature). Cylinder bore measurements were taken as "long" (parallel to crankshaft axis) and "thrust" (perpendicular to the crankshaft axis). Also, three measurements per cylinder were taken: top, middle, and bottom.

Spend a little time examining the accompanying charts. From them you can determine the

350 010									
	COLD READINGS		HOT READINGS			COLD READINGS		HOT READINGS	
CYL No.	Thrust Tenths	Long Tenths	Thrust Tenths	Long Tenths	CYL No.	Thrust Tenths	Long Tenths	Thrust Tenths	Long Tenths
1	5	- 5	38	40	2	24	- 18	58	10
	10	- 15	43	17		31	- 25	66	7
	13	- 10	40	20		28	- 5	60	20
3	5	7	35	42	4	5	14	30	48
	19	- 30	47	5		15	- 5	48	32
	15	- 15	47	15		17	0	48	43
5	5	0	35	32	6	10	- 20	40	28
	13	- 30	42	10		22	- 30	57	5
	12	- 3	48	20		30	- 10	62	15
7	10	0	40	25	8	10	- 5	40	35
	15	- 20	43	27		15	- 13	43	18
	10	- 5	38	25		15	- 5	40	20

Bowtie									
	COLD READINGS		HOT READINGS			COLD READINGS		HOT READINGS	
CYL No.	Thrust Tenths	Long Tenths	Thrust Tenths	Long Tenths	CYL No.	Thrust Tenths	Long Tenths	Thrust Tenths	Long Tenths
1	6	15	41	55	2	2	2 ?		
	9	12	45	50		9	1 ?		
	5	10	40	45		9	1		
3	25	- 2	60	45	4	25	2		
	25	3	60	40		25	1		
	18	6	54	38		19	1		
5	5	5	40	60	6	8	1		
	2	2	38	45		8	5		
	0	0	38	40		2	3		
7	2	2	60	40	8	2	1 ?		
	2	2	60	40		22	1		
	12 ?	3	58	40		22	1		

Dart 060									
	COLD READINGS		HOT READINGS			COLD READINGS		HOT READINGS	
CYL No.	Thrust Tenths	Long Tenths	Thrust Tenths	Long Tenths	CYL No.	Thrust Tenths	Long Tenths	Thrust Tenths	Long Tenths
1	- 5	- 4	32	37	2	- 5	- 5	25	32
	- 4	- 4	32	35		- 5	- 5	30	32
	0	- 3	35	33		- 5	0	30	33
3	- 10	- 3	25	42	4	- 7	- 3	25	40
	- 6	- 3	28	35		- 5	- 2	27	38
	- 3	2	23	32		- 5	0	33	38
5	- 5	- 5	25	40	6	- 5	- 2	25	40
	- 5	- 3	28	38		- 5	- 2	25	35
	- 2	0	32	35		- 3	- 2	30	35
7	- 5	- 2	32	33	8	- 4	- 5	30	30
	- 5	- 3	32	32		- 5	- 3	28	30
	- 3	- 4	33	30		- 5	- 3	30	30

Dart Little M Lightened / cold readings @ 65 Deg / hot readings @ 200 deg TQ Plate temp 195 Deg									
	COLD	COLD TO HOT		HOT		COLD	COLD TO HOT		HOT
CYL No.	Out of round	Change Thrust Tenths	Change Long Tenths	Out of round	CYL No.	Out of round	Change Thrust Tenths	Change Long Tenths	Out of round
1	17	32	40	25	2	15	35	35	15
	15	32	38	21		15	32	35	18
	15	32	35	18		17	34	32	15
3	0	28	44	16	**4**	**0**	**24**	**44**	**20**
	2	30	35	7		**2**	**27**	**38**	**13**
	2	30	30	2		**7**	**35**	**31**	**3**
5	0	30	42	12	6	2	30	42	10
	2	30	38	10		2	27	40	15
	0	30	32	2		5	34	30	1
7	12	32	33	13	8	8	33	38	13
	10	32	33	11		12	32	33	13
	0	27	30	3		6	33	35	8

dimensional differences and accuracy measured between the two methods used: torque plates and hot-honing versus neither. Particularly interesting is the data that shows the number of honing passes required to bring a cylinder into concentricity as depicted in the chart entitled "Dart Little M Lightened" (note cylinder No. 4). This same chart also shows a comparison (same cylinder) of cold readings at 60 and 70 degrees F and one at 70 degrees with the torque plate removed. Here, the two readings make a direct comparison between readings with and without the plate. Remember, all readings are shown in tenths of a thousandth of an inch.

SOME CONCLUDING THOUGHTS

Is hot-honing worthwhile? Some observations may be made from information gathered in this story. For example, with the use of torque plates, stresses and bore distortion created by the installation of cylinder heads can be imposed. Therefore, it's possible to finish-hone cylinders to compensate for such deformities. However, the body of information gathered from hot-honing indicates greater bore distortion is created by circulation of hot water through a block's cooling jacket than by cylinder head clamping forces.

It was also apparent that cylinders in blocks that have been "lightened" will notably benefit from the hot-hone process and should be finish-honed after all material is removed.

Finally, as noted by Wells, bore-gauging cylinders that have been hot-honed and allowed to cool will produce measurements that appear quite skewed. Once they are brought up to temperature, these same cylinders will assume the desired concentricity and, according to the data, be more concentric than if the hot-hone process had not been employed.

Where simulations or approximations are used to approach in-use conditions (in a variety of engine or parts applications), the nearer a process comes to duplicating those conditions the more beneficial it becomes. In terms of cylinder-bore concentricity and preparation, it appears the hot-hone process closely approximates the dimensional environment of a running engine's cylinders. Smokey would likely have approved.

FROM STEPHEN KUZARA'S PERSPECTIVE

Given the improvement in bore concentricity and consistency the hot-hone process enables, what sort of power (or parts durability) can be expected from using the method?

"I'd say less ring land wear, possibly ring wear itself, and ring tension mutation. I believe bore surfaces to be somewhat smoother, resulting in less friction (horsepower). In terms of power, there are no magic numbers. During development of the process, I didn't have access to a dyno facility to verify differences, but the engines just seemed freer to me. One ring company stated that you could expect up to 5 hp in a 600hp engine.

A prominent NASCAR team told Kwik-Way they were seeing about 5 more horsepower after 50 laps than when they were coming off the dyno. At that level of horsepower, the cost was around $100,000 per horsepower. Formula One engines have been hot-honed for years, and several other people using the process swear by it. To date, I've not seen any really good data. Logically, it makes sense. We just know it works."

On the assumption an engine builder feels current methods of finish-honing a block are adequate, what reasons can you provide that might change his mind?

"Experience with hot-honing has taught me several things. First, it's the almost one in every eight bores that you find that distorts more than the average in an unpredictable way that makes you feel so good about finding it. The average distortion is several times that found when using torque plates. In addition, I have found many cracks and leaks even after a block has been closely inspected.

"Hot-honing also reveals a lot about each block and the inconsistencies of the manufacturing and machining process. For example, when you lighten a block, it had better be done before the honing is done. (The data charts show that it required almost 0.008 inch to straighten bores in the Dart Little M block that had previously been lightened.)

"Finally, I believe rings are being launched off the bore surface irregularities, and when they land they begin to bounce, creating destructive harmonics, increased wear, and reduced cylinder pressure. These harmonics may be why some

Dart Little M Lightened / cold readings @ 65 Deg / hot readings @ 200 deg TQ Plate temp 195 Deg									
	COLD READINGS		HOT READINGS			COLD READINGS		HOT READINGS	
CYL No.	Thrust Tenths	Long Tenths	Thrust Tenths	Long Tenths	CYL No.	Thrust Tenths	Long Tenths	Thrust Tenths	Long Tenths
1	- 2	15	30	55	2	0	15	35	50
	0	15	32	53		0	15	32	50
	0	15	32	50		- 2	15	32	47
3	8	8	36	52	**4**	**8**	**8**	**32**	**52**
	8	10	38	45		**8**	**10**	**35**	**48**
	8	10	38	40		**5**	**12**	**40**	**43**
5	10	10	40	52	6	10	8	40	50
	8	10	38	48		8	10	35	50
	10	10	40	42		8	13	42	43
7	5	17	37	50	8	2	10	35	48
	5	15	37	48		3	15	35	48
	10	10	37	40		4	10	37	45

Dart Little M Lightened								
Step by Step Honing of Cyl No. 4 @ 200 Deg						60 Deg/70Deg/TQ plate off		
HOT READINGS			HOT READINGS			COLD READINGS		
No. of hone pass	Thrust Tenths	Long Tenths	No. of stone pass	Thrust Tenths	Long Tenths	CYL No.	Thrust Tenths	Long Tenths
1st	48	60	5th	71	73	4@60	54	37
	55	58		74	75		54	42
	53	57		75	74		50	48
2nd	58	65	6th	77	78	4@70	57	40
	67	65		80	78		58	48
	65	67		79	78		55	53
3rd	63	67				4@70	58	43
	69	68				plate	57	48
	68	68				off	54	53
4th	68	70						
	73	70						
	72	72						

Dart Little M Cyl No. 4							
Cold to Hot Tenths			Change Tenths	Hot to Cold Tenths			Change Tenths
8	to	32	24	77	to	54	23
8	to	35	27	80	to	54	26
5	to	40	35	79	to	50	29
8	to	52	44	78	to	37	41
10	to	48	38	78	to	42	36
12	to	43	31	78	to	48	30

piston ring lands wear out so quickly and, I expect, contribute to decreased ring seal and related possibilities. Imagine the speed of a ring approaching a 0.0007 'hill' on the surface of a bore at 8,000 rpm and the resulting deformation of the remainder of the ring as it tries to accommodate this vandalistic attack."

How cost-effective is this process?

"Besides offering an opportunity to shorten the time required to finish-hone a block, it also doubles as a hot-pressure test. It is also a good method to check out different types of gasket materials and sealants. When I was machining blocks, I charged an extra $150 (1980 dollars) and had no complaints at this price. Except for torque plates, not many pieces of shop equipment will pay for themselves in a dozen uses."

What are common mistakes that should be avoided when using the hot-hone method?

1. Not measuring a bore while it's cold. You lose your frame of reference.
2. Forgetting to make sure all plugs and connections are watertight.
3. Build a water trap in the hone to help keep the water from going through the pump and being mixed with the oil. You will have an accident or two.
4. Venting the oil smoke in some manner.
5. Continually checking and adjusting the gauge. It will pick up heat and expand.
6. Not writing down honing figures. I use the torque plate as a slate.
7. Turning on the heater without any water in the tank. These are very dense heating elements and will burn out in about ten seconds if fired dry.
8. Remember, the system is as hot as an engine fresh off the track.

Are there types of engine blocks that would benefit more than others from the hot-hone process?

"I'd say the following is a pretty comprehensive list: siamese bores, factory blocks, blocks extensively lightened or have other metal removed, welded blocks, sleeved blocks, and block/head combinations with dissimilar metals . . . cast-iron blocks with aluminum cylinder heads obviously need aluminum torque plates."

What is the hot-hone question (and answer) you have found to be the most often asked by an engine builder?

"Without a doubt: 'How do you know what size to hone the bore to?' The answer to this is pretty lengthy, and I'd like to refer readers to the operator's manual that's available from Kwik-Way (www.kwik-way.com)."

Chapter 7

CRANKSHAFT SELECTION

Text and Photos by Jim McFarland

Measured heavy metal (tungsten) round inserts in the crank's throws help to counter torsional vibrations that can affect ultimate torque output. Crankshaft balancing is worth the extra cost and effort to improve its service life and get the best torque application from the engine.

Converting cylinder pressure to crankshaft torque, particularly since these pressures produce intermittent loads, involves a complex set of problems. Too often, it seems we become laced into concerns for crankshaft weight, durability, and balance. Think of this component as a flexible part. As cylinder pressure creates intermittent torsional loads in a crankshaft, there is evidence of continual elasticity within the part—steel begins to act like rubber.

Even though pistons and connecting rods tend to become shock absorbers to these loads, radial deflection (both positive and negative) leads to oscillations created about the crank's axis of rotation. For now, it's important to recognize that crankshafts, as a function of torque produced, tend to become "rubber" in the manner by which reciprocating motion is converted into rotary motion. crankshafts can have influence on such variables as valve timing, intake manifold selection, and header sizing.

WHAT IS TORQUE?

Torque is the result of a force being applied on an object at some distance from its axis of rotation. If we allow torque to act during a period of time, we observe horsepower. Crankshafts make torque. Cylinder pressure causes torque to be made. In the case of engines, for the most part, how much torque is made depends upon cylinder pressure and the distance from crankshaft axis of rotation to the point of applied force—cylinder pressure acting on a piston/rod assembly. Crankshaft stroke plays a part in this factor. In other words, the longer the stroke, the greater the distance from the axis of crank rotation, and the more torque produced. Long-stroke engines make more torque at lower rpm than shorter-stroke ones, right?

Now, while this may appear oversimplified, it's important to understand how a crankshaft reacts to torque and the influence this reaction can have on other engine parts and functions.

CRANK MATERIALS AND CONSTRUCTION

Essentially, the range of crankshaft materials runs as follows: billet steel, steel forgings, cast steel, nodular iron, malleable steel or (in some cases) cast iron. If we were to produce one crankshaft design and reproduce it in all these materials, the order of strength would approximately follow the same list. While cast cranks are typically less expensive than forgings, they can be produced in shapes not available with forgings.

The configuration of fillet radii is often a subject of discussion and frequently believed to be critical in the design, modification, and service life of a crankshaft. If we were to perform a stress analysis test that included all other design features and conditions of a given crankshaft, fillet radii could be considered the most critical factor in overall design and/or modification procedure. There is belief among crankshaft manufacturers that the use of fillets of non-constant radius—sometimes called non-circular contours—is preferred over those of constant radius. This is an area worth discussing with your engine builder or crankshaft manufacturer of choice.

LIGHTWEIGHT VS. HEAVYWEIGHT ROTATING MASS

Let's talk about transient torque. For purposes of this discussion, transient torque is a measure of how quickly an engine can accelerate (including under load) through its

While this illustration depicts a "conventional" style of crankshaft torsional vibration damper, the functional impact of the damping material can be performed by other means that include fluid couplers, friction discs or other load absorbing or damping mechanisms.

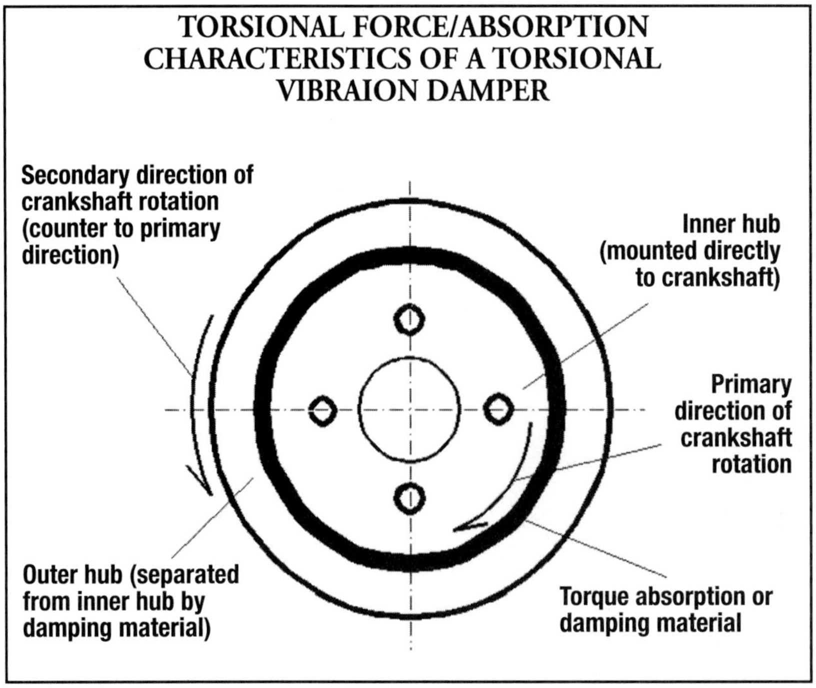

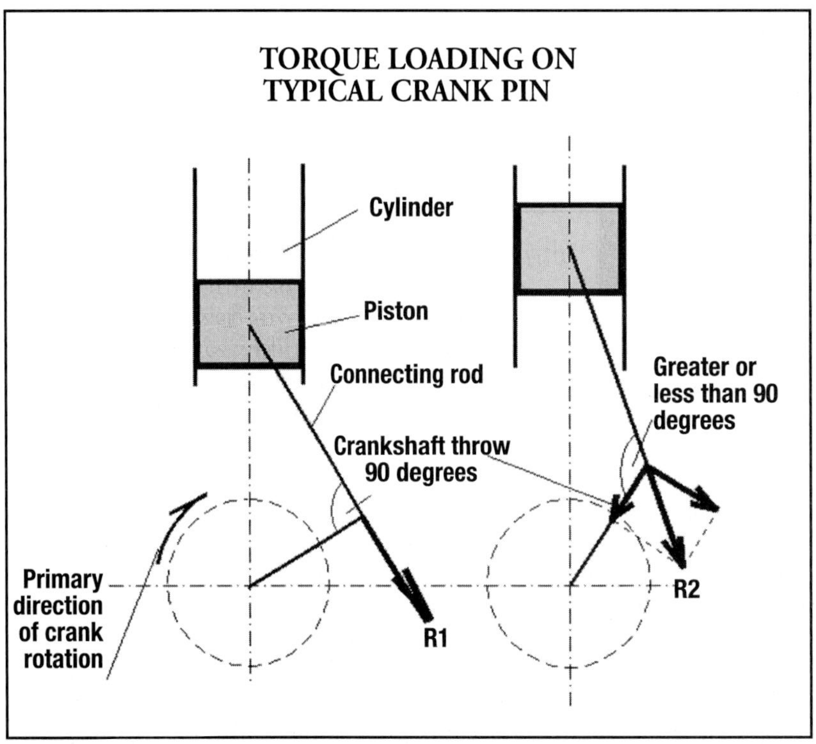

This illustration shows some of the torque load relationships in a typical crank/rod assembly. R1 represents maximum load or torque delivery at an angle of 90 degrees. At angles greater or less than this value, the resultant R2 is composed to two corresponding loads that place forces along the axis of the crankshaft throw. As crank throws are made shorter or longer, piston position at maximum load delivery changes. This is an area mentioned in the story. Companion parts modifications (camshaft, intake, manifold, etc.) can be matched to stroke changes.

useful rpm range. Stated another way, how fast an engine will span from low to high rpm under sudden conditions of WOT. From a measurement standpoint, this is torque as measured on an inertia dyno—not a so-called "accel" test as performed on an engine dyno whereby there is a controlled unloading of the power absorption unit. This is "real-to-the-track" torque, and it relates to an engine's ability to overcome its internal resistance (inertia) to gaining rpm.

Based on these considerations, it is fair to say crankshafts don't normally operate at constant rpm. They're either accelerating or decelerating. Their resistance, in either case, includes static weight and dimensional landscape (stroke length, location and distribution of mass, and so on). Technically speaking, in a dynamic environment, crankshafts are continually changing potential energy into kinetic energy. These are all factors that go right to the issue of how much torque is available at the output end of a crankshaft—they need to be considered for power optimization.

From a practical standpoint, acceleration of a heavy crankshaft absorbs more torque than one of less weight, thereby reducing the amount of net torque available to accelerate the car. But there are trade-offs in terms of durability, flexibility, and potential longevity that should be considered when trimming crankshaft weight. Furthermore, it's not all about weight. Placement of weight, relative to a crank's axis of rotation, is also important. For example, the moment of inertia (resistance to a change in the state of rotation or acceleration and deceleration) increases as weight is moved away from the axis of rotation. Even between two crankshafts of the same total static weight, the one with more weight near its axis of rotation will exhibit less resistance to a change in rotational speed; it has a lower moment of inertia. Keep this in mind when adding heavy metal to crankshaft counterweights during the process of dynamic balancing.

Finally, where total crankshaft weight relates to overall flexibility, it's best to err on the side of stiffness if this may be accomplished by selecting a crank that tends toward stiffness in combination with durability and light weight.

TORSIONAL VIBRATION AND DAMPING

Visualize a single-cylinder engine running in very slow motion. Upon a rapid increase in cylinder pressure, quick application of force is applied to the crankshaft. The throw connected to the piston/rod assembly is deflected in the direction of crank rotation. Stated another way, it gets a "kick in the butt" that amounts to an increase in crankshaft rotational acceleration. In so doing, the crankshaft's throw acts somewhat like a spring in that once its maximum deflection is reached, elasticity of the crankshaft material causes the throw to act against the load that caused its acceleration increase.

Until the next firing cycle and subsequent load on the crankshaft throw, this action and reaction condition continues as an oscillation of the crankshaft throw, even though the crank is in rotational motion. Note the accompanying illustration. If you were to view this example crankshaft along its axis of rotation, the oscillation described would appear as clockwise and counter-clockwise movements of the crank throw, all during normal rotation of the shaft.

Now, transfer this image to a multi-cylinder engine. Since the delivery of each firing impulse causes its respective crankshaft throw to experience torsional oscillation (or vibration), imagine the interaction of these pulses along the length of a crank and you begin to realize the complexity of the overall system. Moreover, since all the pistons are tied to this oscillating crankshaft, interruptions in smooth piston movement result. Plus, if the engine has a carburetor, we have now linked a torsionally vibrating crankshaft to fuel delivery and the corresponding effect this can have on carburetor calibration and power.

Further, if piston motion has been interrupted and is no longer directly coordinated with specific crankshaft angularity position, is it not reasonable to assume the relationship between valve timing and piston position has been upset? Well, it can be. Notwithstanding the fact that camshaft timing chains or gear sets become shock absorbers to the torsional oscillation of crankshafts, it's possible to address this problem by the proper selection and use of torsional oscillation (vibration) dampers.

CRANKSHAFT THROW DEFLECTION FROM TORSIONAL VIBRATION

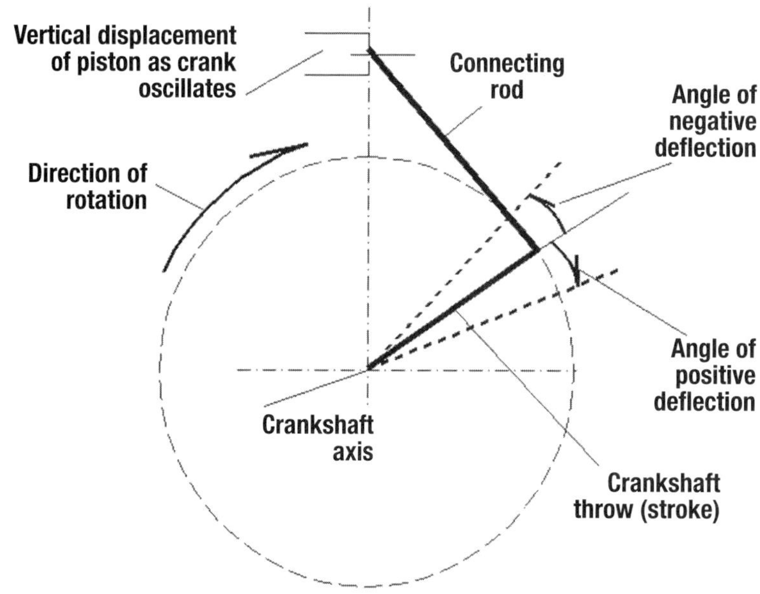

Use of torsional vibration dampers is intended to minimize crankshaft oscillation throughout rpm range. Negative torque values can approach 8 percent of positive values.

TUNING CRANKSHAFTS

What happens to net torque when a crankshaft deflects against its normal direction of rotation? Power is lost—plain and simple. How much power, and where in the rpm range does this occur? Well, there are ample variables involved, and the rules of thumb recognize that the amplitude and rpm at which maximum and minimum deflections (and power losses) occur varies and include the following: the level of power produced, range of rpm, dynamic properties of the engine's rotating and reciprocating assembly, static weight of the moving parts, and other factors.

Based on orders of critical engine speed, there will be certain rpm at which negative torque (counter-rotational crankshaft deflection) is greater than at other rpm. For example, negative torque measured at 3,500 rpm may be less at 4,000 and then more at 4,500.

Specifically, in an engine producing 540 shaft horsepower, I've seen negative torque losses approaching 35–40 lb-ft of torque, in an otherwise torsionally damped engine. Other examples have been greater and less. The point is that such losses can be neither insignificant nor unworthy of elimination or reduction.

How Individual Cylinder Torque May Vary as a Function of Crankshaft Throw Oscillation

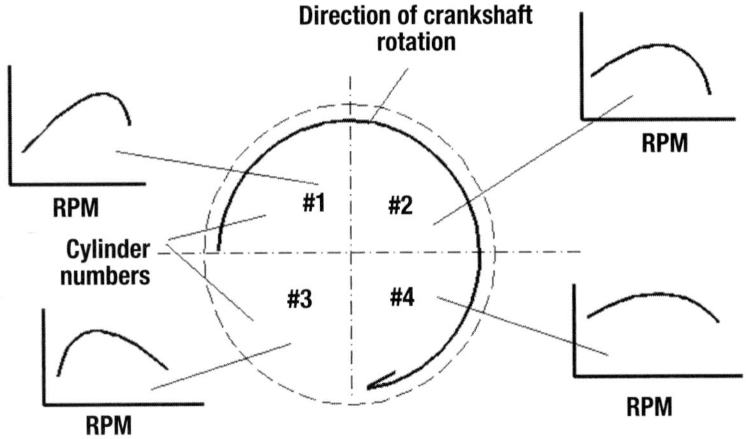

Cylinders shown are not in any given firing order. The purpose of this illustration is to show how cylinder-to-cylinder torque output can vary as a function of crankshaft deflection. Graphs are only to illlustrate the point. In reality, specific torque values can be determined by several methods, including the use of Engine Cycle Analysis where individual cylinder performance is measured by in-cylinder pressure analysis.

Because the engine speed and extent of power loss vary, can crankshafts and dampers be tuned to these rpm (or frequencies)? The answer is yes. From a practical standpoint, damper tuning is often more easily accomplished than crank tuning—unless, of course, you're an accomplished engine balancer. Know the range of rpm where the engine will spend most of its on-track time and then target the damper selection accordingly. If you have difficulty obtaining this type of information from your choice of damper manufacturer, be persistent. You will likely come away from the experience with the following: 1) the correct part, 2) an urging that this is unimportant information, 3) it can't be done, or 4) don't split horsepower hairs to this extent. If you're trying to increase parts life and make more power, opt to accomplish point #1.

CRANKSHAFTS AND WINDAGE

Even the use of dry-sump oiling systems does not prevent crankshafts from being influenced by the migration of engine oil from the crankcase to the pan. It's a given that oil will become attached to a spinning crankshaft, adding to its dynamic weight and inertial resistance. In fact, high-speed photography reveals that the oil can look like toffee rope wrapped around the spinning crankshaft. By the use of various oil pan devices, attached oil can be physically removed from crankshaft surfaces. Since tapered edges tend to release/shed oil more easily than "squared off" portions, the shaping of counterweights is another common method of addressing the problem. Special coatings of the type that reduce surface tension, applied to crank throws and counterweights, can also be used. But in any event, assume that oil passing by the crankshaft and pulses of air created by rapidly descending pistons and the volume of air entrapped inside them will combine to increase the importance of reducing the amount of oil that becomes a "passenger" at high crank rpm.

CRANK'S INFLUENCE ON PARTS

Let's say you decide to increase crankshaft stroke. This implies an increase in low- and mid-rpm torque is desirable as well as the selection of an intake manifold, camshaft, and header system that favors torque output in the same range of engine speed. Similar to events from lengthening connecting rods (all else being equal), a stroke increase changes the rate at which intake flow velocities are created (versus crank angle). It also affects piston dwell around TDC and BDC. This suggests an adjustment of sparking timing (at least initial spark) when comparing engine applications of a stroked and unstroked crank.

Maybe you've decided to decrease the inertial resistance of the crank you're using. That suggests an engine that will accelerate quicker and, likely, one that would benefit from an increase in compression ratio. Short of performing single-cylinder dynamic pressure analysis as a function of crank angle (perhaps with Engine Cycle Analysis or comparable in-cylinder measurements), it's possible to recoup some power lost to retarded rear cylinders by juggling rocker ratios that serve them. Consider what you'd do to improve higher rpm power with valve timing that is retarded—maybe have a ratio that is a shade higher for an initial test.

It's safe to assume, although you may not know the exact amount of crankshaft deflection (front to rear), that an engine's rear cylinders are running slower than those in the front. Depending upon how much deflection is occurring, rearward cylinders may be running in a retarded position relative to those in front. Some adjustment to spark timing to these cylinders can be helpful. Camshafts also deflect, so this condition of retard can result from a stacking of events from deflection in both cam and crank. To help minimize this problem, reducing the amount (frequency and amplitude) of crankshaft deflection by the sensible use of a torsional vibration damper is mandatory to extending parts life and optimizing power.

CRANKSHAFT SELECTION

Dennis Wells of Wells Racing Engines often brings practical and valuable insight to the weekly circle tracker. I asked him to expand on points about crankshaft selection.

Why do you think crankshaft balancing is important?

"Clearly, engines need to be balanced during the build-up process. There are dynamic forces that exist in a running engine that probably exceed what many of us might think. Here's an example of forces applied on the crankshaft. At 4 inches from the crankshaft's axis and with 28 grams of unbalance, there's 112 pounds of force at 4,000 rpm. This increases to 448 pounds at 8,000 rpm!"

How important do you consider crankshaft dampers, and why?

"Crankshaft torsional vibration causes crankshaft and flexplate failure, loosened bolts, broken timing chains, and erratic valvetrain and ignition problems. We tested our USAC Silver Crown car on the wheel dyno. We removed an ATI lightweight damper, installed a small aluminum hub, and made five consecutive runs. We saw no horsepower gains. What we discovered when removing the aluminum hub was that the torsional vibration, in only five runs, was so great that the hub was no longer a press fit. It slid off by hand! How much more proof could you want?"

What is your personal choice for torsional dampers, and why?

"The ATI Super Damper. I prefer this unit because you can replace any part that might become worn. They are available with aluminum or steel hubs. We run a steel hub on Silver Crown engines because we crank the engine off the front of the balancer. I have never had a failure out of an ATI damper."

In your opinion, what is the most important consideration when selecting a crankshaft?

"The position of the counterweights, the 'clocking' of the oil holes, the quality of the material, and the distance of the weight from the centerline."

Do you have a preference in the type and shape of the fillet radii between bearing surfaces (rods and mains) and flanks?

"I prefer a 0.125-inch radius. Just about all good racing crankshafts have this radius."

What's on your list when it comes to crankshaft installation and maintenance?

"Always inspect a crank closely before performing any machine work. This includes checking straightness, thrust surfaces, checking every journal for taper, size and roundness, confirming the stroke, and passing Magnaflux inspection. Remember, everybody can make a mistake; you just need to catch 'em. It's easier to correct any of these problems before you waste your time balancing, or worse, building the engine and then finding out something is wrong. You must use chamfered bearings on cranks with a radius larger than 0.062 inch. you must 'mic' everything and check every bearing housing for size and roundness."

Any more relevant comments on cranks?

"We do not remove oil pans and only replace the bearings. From years of experience, I've determined this is a bad habit. The reason is bearing housing bores distort to an out-of-round shape. The used bearing distorts with the housing bore, and you have no problem. Then you install a new, round bearing, which doesn't match the distorted housing bore. The new bearing will come in contact with the crankshaft, and you now have an engine failure. Your first thought might be to blame the engine builder or the bearing manufacturer. I've seen a lot of spun bearings in 34 years of building engines, and they are not pretty!"

Chapter 8

ROD POWER

This Geometric Link Plays A Greater Power Role Than You Think

Text and Illustrations by Jim McFarland

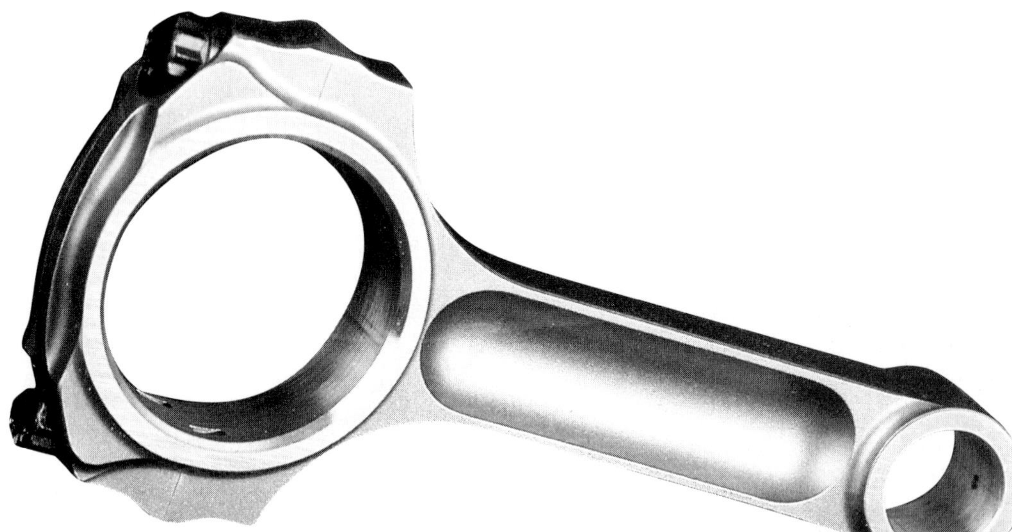

This chapter is about the influence connecting rods have on the ability of an engine to make power, specifically at certain points in its rpm range. It does not contemplate materials, strengths, weaknesses, elasticity properties, or other topics typical of many editorials about rods. Instead, we'll look into the essential relationships of how piston motion, as affected by connecting rod length, relates to valve timing, spark timing, intake and exhaust port sizing, and other elements pertinent to optimizing volumetric efficiency (VE) or torque.

THE "SLIDER-CRANK" MECHANISM

In the study of mechanisms and their dynamics, an assembly common to the internal combustion engine is the slider-crank mechanism. It consists of a rotating crank, a "link" or piston, and a rod connecting the two (see illustration). Since the crank's axis is fixed at one point, the piston can move along a path either on or near a line intersecting the crank's axis of rotation. If the piston's motion is limited to travel only along this axis, then one example of this movement is the traversing of an engine's pistons in its cylinders.

Several mathematical equations describe the motion of pistons as a function of the crankshaft angle. Obviously, since there is a change of piston direction at each end of its stroke, both velocity and acceleration will be zero at these points. However, we are concerned about piston velocity and acceleration between top dead center (TDC) and bottom dead center (BDC).

By one definition, average velocity is a measure of distance traveled in a period of time. Instantaneous velocity, on the other hand, is a measure of velocity at a particular location or point in time relative to that location. More important to our discussion is the definition of acceleration: the "rate of change in velocity" or, depending upon how quickly velocity is increasing or decreasing, how acceleration can be either positive or negative.

As the length of a rod is changed, so are the velocity and acceleration characteristics of its attached piston at locations other than TDC and BDC.

PISTON MOTION AND ITS EFFECTS ON VOLUMETRIC EFFICIENCY

Especially in the case of normally aspirated engines, piston velocity and acceleration impact the rate at which a difference is created between atmospheric and cylinder pressure. It is important to understand that a condition in which cylinder pressure is lower than atmospheric pressure does not cause an engine to suck air. Rather, atmospheric pressure pushes air into the engine. The faster a piston moves downward, the more rapid this pressure difference is created and the quicker atmospheric force can fill the void. That's fundamental.

More to the point, since a piston's motion is a combination of acceleration, deceleration, and two changes of direction (TDC and BDC), it becomes crucial to determine its position in the cylinder bore when the highest rate of pressure drop is being created. Since the piston position in a bore can be related to the crankshaft angle, a fairly straightforward mathematical analysis provides this information.

Now, to the VE issue. Critical to optimizing volumetric efficiency is the timing of valve action to piston motion. Valve timing and piston motion are inexorably tied together and should be considered as such—especially when selecting camshafts and matching them to connecting rod lengths ("short" versus "long").

Illustration of Basic "Slider-Crank" Mechanism Characterizing Piston, Connecting Rod and Crankshaft Dimensional Relationships & Positions

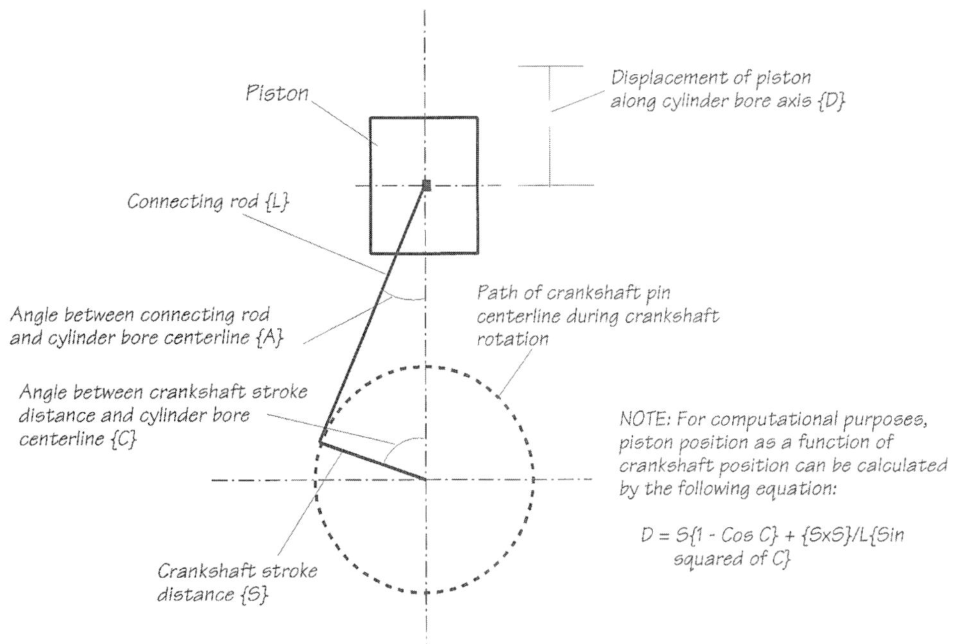

HOW ROD LENGTH AFFECTS PISTON MOTION

Since pistons momentarily stop at TDC and BDC, any change to rod length affects acceleration and velocity between these two end points. Right here, it may be helpful to make a general statement about how changes in rod length affect piston motion.

As a rule, lengthening a rod tends to cause a piston to remain longer (increased crankshaft angle) around TDC and BDC than with rods of shorter length. Let's assume an example in which the stroke distance is the same for two different lengths of connecting rods. In this case, we know each piston will travel the same TDC-to-BDC distance, and the longer rod will cause increased piston dwell time at these two end points.

Let's examine what goes on in between TDC and BDC by cutting right to the heart of the issue. We know the piston for each rod length will achieve maximum velocity and acceleration somewhere during the stroke. What's important is that this point (crankshaft angle) will be different for each rod length. As a piston approaches or departs TDC or BDC, thc longer the rod, the slower this motion will be. As you would expect, the corollary is true for rods of shorter length.

In general, peak torque is largely influenced by the tuning characteristics (particularly dimensions) of the intake and exhaust system. Changes in connecting rod length tend to shift peak power rpm closer to or further from peak torque rpm. For example, as the connecting rod length is increased, the peak horsepower rpm will move toward the peak torque rpm.

PISTON MOTION AND VALVE TIMING

Let's start this off with an example. Suppose we do the math for a slider-crank mechanism dimensioned with the stroke and rod length we'll use and attempt to select a matching camshaft. The numbers tell us the piston is moving at maximum velocity at 75 crankshaft degrees past TDC on the intake stroke. From another perspective, this tells us we have the potential for the highest VE (as affected by piston motion) at this crank angle. If the intake valve opening at this same crank angle is insufficient (or excessive) to maximize inlet flow, we've lost some advantage to optimize VE and torque (recall that an engine's volumetric efficiency and torque potential are virtually synonymous).

Some dyed-in-the-wool engine builders will make graphical plots (on an X–Y coordinate system) that show piston position as a function of crankshaft angle. Then they will superimpose a similar plot of valve-opening/closing dimensions onto the piston position graph. By comparing the two, a graphical representation of how the piston position compares to the valve opening reveals this relationship and what must be done to valve timing for proper matching of the two.

Illustration Showing Basic Connecting Rod Terminology and Areas Frequently Associated With Mechanical Failure

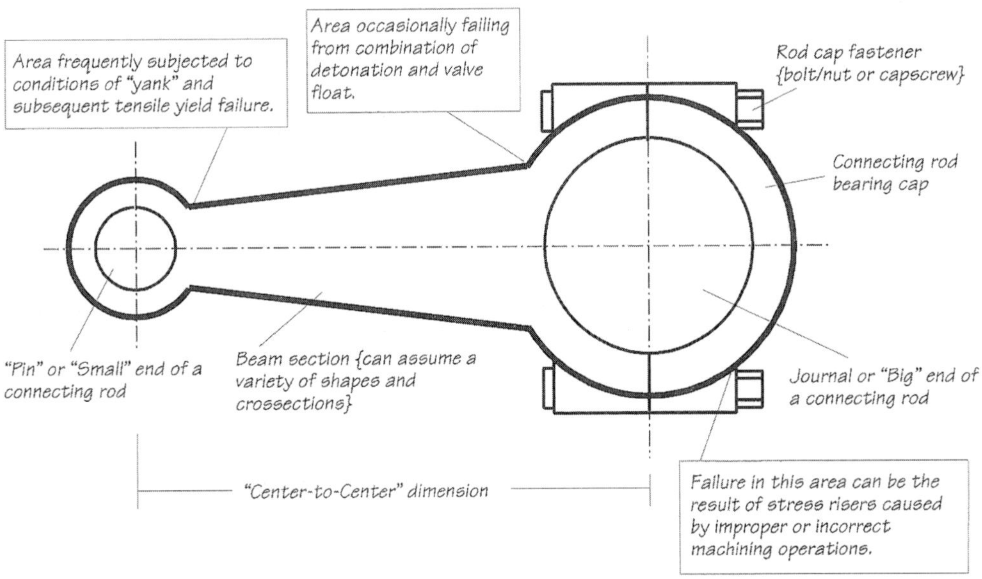

Other engine builders blessed with engine cycle analysis capability will make real-time cylinder pressure versus crankshaft angle traces and do a dynamic analysis of what is required for optimum valve motion to match existing piston motion. Still others will use computer software to perform iterative changes to valve timing and arrive at the same or similar conclusions. By whichever method you choose, integrating valve motion with piston motion is a contemporary must.

PISTON MOTION AND PORT FLOW RATES

Simply stated, if the piston isn't moving very quickly during the inlet cycle, something else must be added that will increase port flow rate. Although this condition should not exist, it often does if proper matching is not accomplished between piston motion and valve timing.

One of these is the use of an intake manifold with small runners (in terms of cross-section area). Small-diameter headers tend to "crutch" in the same fashion. You might even consider advancing the cam as a further means of gaining lost torque from slow piston speeds around TDC (particularly during early phases of the inlet cycle).

Let's say you're not using a long rod. In this case, piston speed at and around the time of the inlet cycle will be quicker, allowing the use of slightly larger porting, manifolds and headers.

Once again, powerful and efficient racing engines are a material integration of interrelated components, and by failing any of the contemporary—and somewhat exotic—methods for making these determinations, simplified rules of thumb are not without value.

PISTON MOTION, CYLINDER PRESSURE, AND IGNITION SPARK TIMING

If you are doubtful that these are related, consider this. If a piston stays around TDC longer, it'll do more work on the crank, perhaps, but let's consider something that goes along with this.

There's a term in the study of internal combustion engines called constant volume combustion. This amounts to a set of conditions that unravels rapidly throughout the burn while the piston is changing direction (stopped) at TDC. At this point, cylinder pressure and temperature rise suddenly and to high levels, sometimes tending to cause detonation or uncontrolled combustion.

As connecting rod length is increased, instantaneous cylinder pressure becomes higher (all else being equal) approaching and at TDC than with a shorter rod. As a result, it is often necessary to run less initial spark timing with increased connecting rod length.

If we consider the amount of work done on a piston prior to TDC (negative) and after TDC (positive), it helps net power to reduce the amount of negative work. Net work (the difference between positive and negative) is called Indicated Mean Effective Pressure (IMEP). You could also say this is what we're calling "net work." While maintaining piston residence in the vicinity of TDC, the idea is to increase the dwell time for pressure on the piston, just past TDC. This increased pressure (from a longer rod), which also increases pre-TDC mixture density and flame rate, suggests the possibility of slightly less mechanical compression ratio (to avoid detonation) and less initial spark timing, all of which raises the level of IMEP or net work output.

MORE ABOUT CONNECTING RODS

In a sense, rods are also shock absorbers. Considered in slow motion, combustion pressure development is not sudden. If proper, it builds in a uniform fashion that includes rate of flame travel, mechanical loading, and related combustion-pressure conditions. Detonation, on the other hand, is akin to the force of a hammer striking a piston top. While pre-ignition can lead to detonation, the ultimate effect on connecting rods is the same. All these forces are ultimately transferred into the connecting rod, comprising multidirectional stress loads, many of which lie outside the geometrical axis of the rod. Part of this is caused by asymmetrical forces applied to the tops of pistons, the other part by dynamic action of the rods as connected to a crankshaft. Add to this the fact that "load sharing" is occurring among connecting rod sets, and it's possible for detonation in one cylinder to cause damage to a rod not associated with the problem cylinder.

It's possible to design a connecting rod that is too stiff. This can allow combustion pressure to transfer too rapidly to bearings (rod and crank) and an inability to absorb unavoidable pressure spikes associated with abnormal combustion. Furthermore, connecting rod stiffness comes into play when it's necessary to absorb torsional loads caused by rotary vibrations in a piston—somewhat like wringing out a wet rag.

Sketch of Simplified Piston Motion Showing Points of Maximum, Minimum Acceleration and Velocity

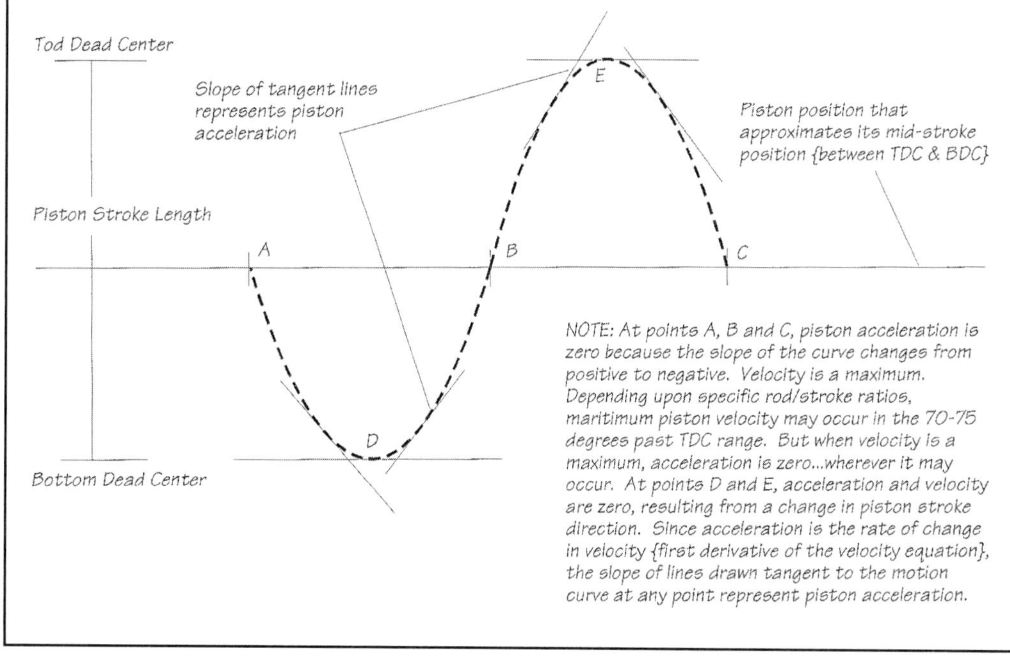

MATERIAL SELECTION

Important to this decision is the ultimate use. Of the typical materials used, the following are common: cast iron, steel, titanium, aluminum, and powdered metal. Produced by methods that include casting, forging, or high-pressure die techniques, end use is the primary consideration for manufacturing choice—excluding economic factors. As is the case of virtually any engine component, especially for high performance or racing, a trade-off between cost and durability is unavoidable.

Interestingly, there appears to be a merging of economics in the production of aluminum versus steel connecting rods. Contemporary design methods of each, which include computational stress analysis of designs, seem to be reducing the historical differences between aluminum and steel connecting rods. Despite this convergence of choice, steel rods remain higher in long-term durability than aluminum, again depending upon application.

At least in theory, if not practice, high-impact loads upon connecting rods seem to favor aluminum in comparison to steel counterparts that may tend to burst under the same type of loads. What must also be considered is that the most damaging loads on a connecting rod do not occur during the pure compression of combustion. The oscillatory vibration of a crankshaft causes connecting rods to experience compression and tensile loads, alternatively, during every rotation of the crank. Essentially, they become rubber bands that must withstand these cyclical load patterns while maintaining sufficient dimensional integrity to not fail or cause piston/head contact or bearing damage, which brings us to a final area of importance.

CAUSES OF FREQUENT ROD FAILURE OR DAMAGE

Consider the comments of Rocky Childs (Childs & Albert): "There is a consistent opinion among many of the engine builders who are our customers that rod failure is most often caused by excessive

Illustration Showing Some of the Effects From Changing Connecting Rod Length

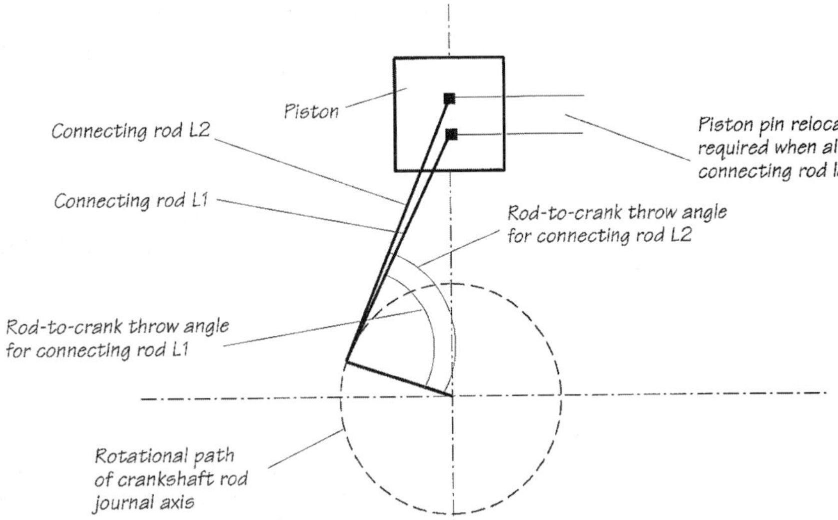

NOTE: Several observations can be made by studying this illustration. In particular, note the apparent mechanical advantage afforded by the longer rod {L2} in applying combustion pressure forces onto the crankshaft journal {as compared to rod L1}. Additionally, "dwell time" of the longer rod in approaching and departing TDC is increased, allowing an extended time for torque application and faster burn rate from increased effective mixture density. Of the results this creates {longer rod}, peak horsepower is developed sooner in relation to peak torque rpm, causing less rpm to separate peak power and peak torque...ideal for engines operated in a relatively narrow span of rpm {superspeedway, etc.}.

combustion or detonation pressure. This simply isn't so. Now, that statement doesn't mean connecting rods won't fail during these times, but it's not the predominant period. In our experience, and I'd estimate about 90 percent of the cases we've specifically evaluated, rods don't fail during compressive loads. Actually, our data strongly indicates it's during the exhaust stroke that a connecting rod gets yanked apart as it departs from TDC.

"This isn't our imagination. If you look closely at the appearance of material along the parting surfaces between the failed pieces, you can visually note how the material has failed under a tensile load. This, of course, assumes damage to the parts that followed; the failure doesn't mask these areas and cause them to be unreadable. Furthermore, failures of this type usually occur just below the pin area."

Two other areas are commonly associated with rod failure. "Valve float, sometimes not even recognizable by a driver or engine builder because it's not always audible, takes away the cushion provided by combustion pressure as the pistons approach TDC. As a result, there's nothing to soften their approach to cylinder head surfaces, and pistons get thrown against the heads. It doesn't take much imagination to visualize the forces present when a high-revving piston passes through TDC with little or no resistance. It's not just about valvespring damage. Other factors come immediately into play, too."

Finally, there is another area of concern regarding rod failure. "Right where a connecting rod's beam section changes in area from wide to narrow, we find frequent failures. At C&A, we call this the 'hinge point,' an area to which we devote a lot of time determining the best dimensional compromises compared to rod material selection. It's something that should be discussed with any manufacturer when making a rod set purchase," says Childs.

BOTTOM LINE?

Connecting rods are not just objects that tie pistons to crankshafts. Aside from structural considerations, their length affects piston motion versus crankshaft angle, which is linked to valve timing and rpm and is integral to an engine's volumetric potential and torque output.

Chapter 9
PISTON TOPS
Flooring The Combustion Space

Text and Illustrations by Jim McFarland

This "flat-dish" design displays relatively uniform burn throughout the central portion of the piston's top. The darkened areas suggest richer, slower burning portions of the total mixture associated with crevice volume (the space around the piston circumference and above the top ring).

If you often build, rebuild, or are just plain curious about optimizing the power from your engine, this subject is required reading—and critical to producing race-winning horsepower. Simply stated, it's just that simple.

Here's the deal. By improving upon the homogenization (mixing efficiency) of fuel and air prior to and during combustion, it's possible to increase flame speed. As the speed of combustion is increased, the amount of ignition-spark timing (before top-dead-center, or TDC, piston position) required to make the burn can be decreased. Fast-burn engines like reduced spark timing.

The later the point of ignition, the less work (pressure) is applied against the piston's motion approaching TDC during the compression stroke (if flame speed could be increased to the point of instant complete burn, the ideal would be to introduce the spark at the moment of TDC). And as this "negative" work is reduced, the net amount of work (pressure) applied to the piston as it moves down during the power stroke is increased. If you equate this pressure (before and after TDC) to torque, which you can, it should not be difficult to recognize the benefits of increased flame rate.

Of the ways mixing efficiency can be improved, so-called mixture motion is currently undergoing study and development by some of the finest engine designers and builders in the business—including the OEM motorsports community. This story explores one facet of mixture motion as applied to the layout of or modification to piston "crowns" or tops.

IN ITS EARLIER STAGES

Sir Harry R. Ricardo made certain statements about turbulence in his academically classic High Speed Internal Combustion Engine books of the early 1900s. Among them was the following: "The rate of burning depends primarily upon the degree of turbulence and may be expressed in terms of increase of pressure per degree of crank angle." This approach is currently used in the method of Engine Cycle Analysis (ECA) being employed by leading Nextel Cup teams. His statement clearly defines the fact that mixture completeness (homogeneity) and burn rate are related.

Ricardo also noted that the need for turbulence decreases in proportion to an increase in mechanical compression ratio (c.r.). Stated another way, as compression ratio is lowered, the need for and benefits from mixture motion increase.

For example, gains from turbulence in a 9:1 compression ratio engine gasoline will be more substantial than in one of 13:1 compression ratio, although some gains are possible at the higher compression ratio. In addition, Ricardo suggested that a turbulent vs. non-turbulent engine tends to experience detonation earlier by comparison. This provides an opportunity to remove a measure of spark timing, allowing for an increase in net cylinder pressure after TDC and subsequent gain in torque (positive vs. negative work).

BASIC IN-CYLINDER MIXTURE MOTION

Of the types of motion popularly being studied, swirl and tumble are foremost. Motion that tends to be directed along the plane of inlet flow generally describes swirl. This can be movement in the same direction as this flow, or in an opposite direction. Positive swirl is motion in the same direction as incoming flow. Negative swirl is in an opposite direction but still in the plane of the inlet flow. Tumble is motion akin to rolling a ball down a flight of stairs—essentially end-over-end flow in a plane roughly

Illustration of Boundary Layer Construction & Eddy Formation

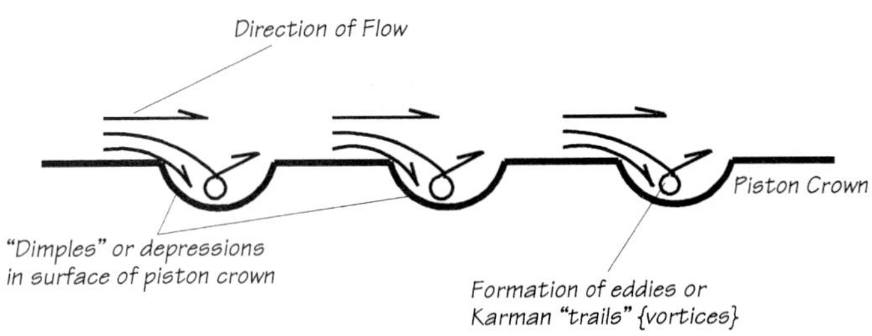

NOTE: Particularly in areas where fuel tends to separate from the air and activity along such areas is low, the addition of dimples helps "activate" the boundary layer region, thereby aiding fuel to be re-suspended AND broken into smaller droplets. Once re-mixed as smaller particles, the resulting mixtures tend to burn faster, providing an opportunity to reduce net spark timing, increase net torque, thereby improving throttle response and on-track fuel economy.

AutoCom 8/5/00

Hi-Tech's Swirl Quench design technology has become quite popular. It utilizes a tapered ramp to aid in mixture-motion direction and dimples in areas of low boundary layer activity to improve mixture quality (homogeneity). Data from testing this concept includes lower exhaust gas temperatures, reduced spark timing requirements, and higher net-torque output.

Schematic Illustration of Karman "Vortex" or "Ring" Generation

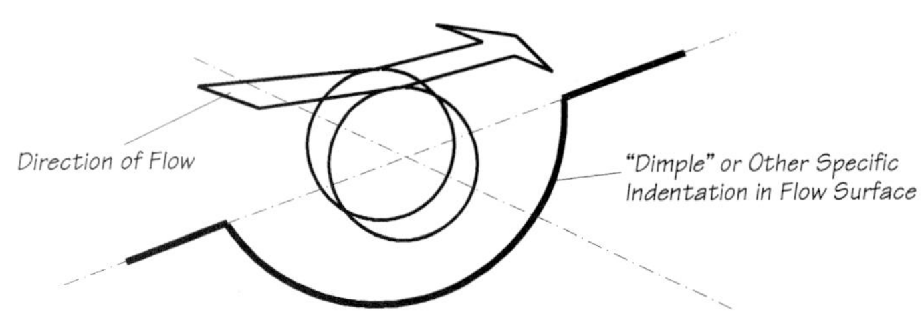

NOTE: While this illustration does not indicate specific re-direction of streamlines in the flow, the notion of boundary layer "tumble" or expansion into the indentation creates energy available to help maintain or return to suspension any separated fuel. The net effect is toward improved mixture efficiency {homogeneity} and higher burn rate, based on smaller fuel particles and an effectively "leaner "mixture.

AutoCom 8/5/00

perpendicular to the swirl plane or axis of the crankshaft.

The shape of a piston's crown can affect both swirl and tumble. In its simplest form, a flat-top piston can influence swirl more than tumble. This is because protrusions above the piston's deck surface are minimized, leaving a broad surface upon which features can be built for swirl enhancement. These include tapered ramps and/or strategically placed texturing or dimples, all intended to modify the quality of swirl, improve quench, and aid combustion efficiency.

Where engine-building rules permit higher mechanical compression ratios, the effectiveness of piston-crown protrusions in the combustion chamber should be carefully considered. Protrusions, or shapes built into the cylinder head, can encourage flame migration toward the exhaust valve, but at the same time they increase compression ratio. This opens the possibility that such protrusions can cause an impediment to flame movement and a disruption of the air/fuel mixture during the burn. The potential result is lost power and possible detonation from excessively lean blends in affected areas.

When considering the advantages of manipulating the shape and texture of the piston crown to enhance in-cylinder mixture motion, remember that there are advantages and disadvantages to swirl and tumble. Even though swirl tends to increase flame rate and help deliver a more uniform rate of combustion, used improperly it can lead to increased spark timing and reduced volumetric efficiency. Excessive swirl can also cause heavier fuel droplets to be centrifuged from the airstream, which disrupts the in-cylinder air/fuel mixture ratios.

Tumble also has advantages and disadvantages. Much like swirl, it can aid combustion rate and uniformity. It can also reduce net volumetric efficiency, leading to losses in net torque. But implemented correctly, either by itself or in conjunction with swirl, tumble can produce increased power.

Today, the instrumentation technology that allows swirl and tumble to be measured is becoming

Here's an example of poor mixture quality. The burn pattern is sparse, especially in the spark plug area. Reduced power is supported by increased spark timing requirements and a loss of throttle response, especially off the corner and when passing.

Conceptual Illustration of Mixture Motion {Intake toward Exhaust}

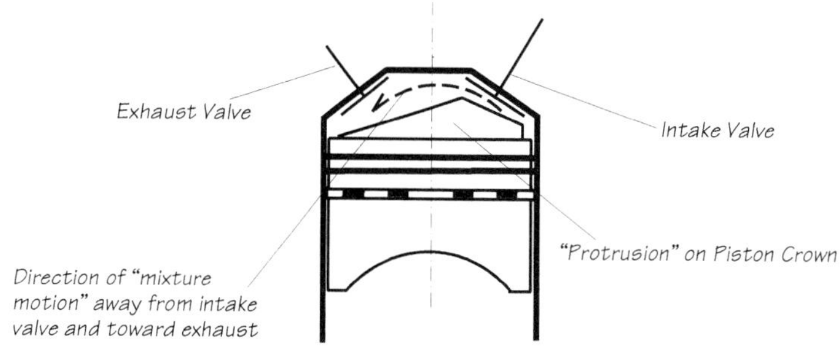

NOTE: It is important to note that the crown shape shown is only indicative of the importance in "driving" the burn away from the intake and toward the exhaust. Various methods are employed to create this effect. Including swirl. One is the Hi-Tech "Swirl-Quench" design {shown pictorially elsewhere} that encourages mixture swirl and incorporates dimples for mixture conditioning. Other methods include "tight" areas between crown and intake valve with a progressively open portion extending toward the exhaust valve. A quasi-quench condition is created between combustion chamber roof and piston crown, concentrating the burn pressure more in the area of the exhaust valve. Typically, a successful result of this method enables reduced spark timing.

AutoCom 8/5/00

increasingly affordable to engine builders and designers. Once priced in a range typical of laboratory-grade equipment, swirl-and-tumble meters are now being marketed for aftermarket and enthusiast use.

UNDERSTANDING PISTON-CROWN BURN PATTERNS

Unfortunately, this is not an exact science. Experience and the correlation of visual inspection with engine dynamometer data (power and brake-specific fuel-consumption information) and on-track performance is currently of high value. However, there are some Saturday-night rules of thumb worth mentioning.

Uniformity of color is one area to consider. We'd like for the air/fuel mixture to be relatively uniform throughout the combustion space (in reality, this is difficult at best); a uniform burn-residue color across the face of the piston is the target. Variations in combustion residue point to different air/fuel ratios that existed during the burn. Dark areas suggest fuel-rich burns, while lighter patterns indicate more air-rich mixtures were involved. Portions not showing color usually indicate a lack of combustion or periods of exposure too brief to leave any residue. These also can point to problems of fuel wash or an overabundance of fuel to prevent complete combustion.

Keep in mind that mechanical compression ratio alone does not always lead to detonation. Poor mixture quality (a wide range of air/fuel ratios within the combustion space) often causes detonation. The ability to achieve and sustain high cylinder pressure must include well-mixed air/fuel charges in order to prevent or limit detonation.

If the engine you're tuning likes spark advance, does not live up to torque-output expectations, requires more fuel than you think necessary, and is detonation-prone, it's a good idea to examine your piston crowns. Each (and all) of these symptoms often is an indication of poor mixture quality, typified by excessively lean and rich regions within the combustion space. This is often the result of a separation of the air and fuel. So spend some time reading the piston tops; it's just another tool for hearing what your engine is telling you. Note the accompanying photos for some examples of combustion problems.

Conceptual Illustration Showing Basic Direction of "Swirl" Mixture Motion

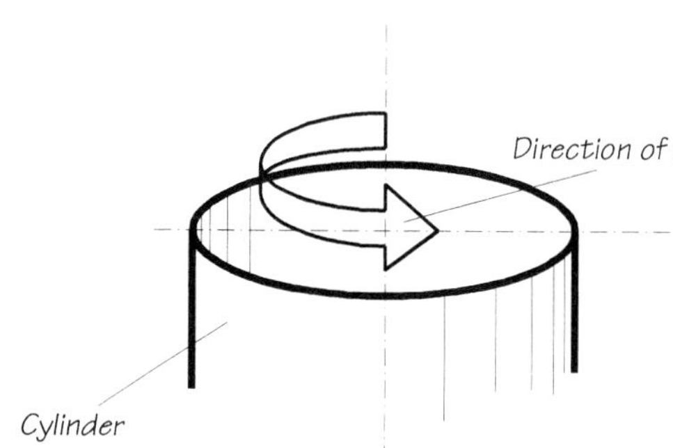

NOTE: Mixture motion can be induced by intake manifold runner, port design, piston crown shape of combinations of all these. In particular, piston crown configuration {especially during the compression stroke and initial portion of the burn} can aid air/fuel mixing and enhance combustion. Tapered pads that direct mixture circumferentially around cylinder perimeter, combustion chambers that enhance this movement or combinations of both can be used to aid combustion.

AutoCom 8/4/00

As viewed in the photo, mixture motion is counter-clockwise. Quench areas (around the perimeter) show little or no combustion activity. Note the evidence of mechanical separation (the cleaned portion) of air and fuel just past the intake valve pad. The sharp edge on the plug-side pad tends to encourage this separation.

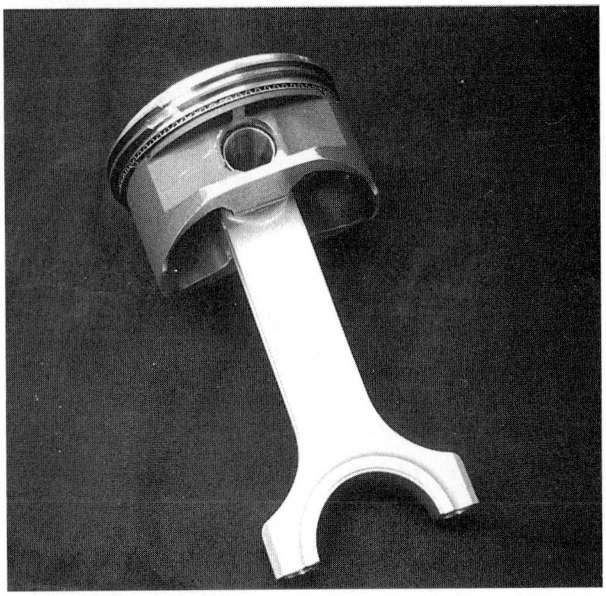

Even though the complete piston is a complex part with almost every dimension held to the highest tolerances, a static area—the top surface—may prove to be the most critical of all.

It is also possible that unequal cylinder-to-cylinder air quantity and quality can cause torque output variations beyween cylinders. Just because all cylinders are receiving equal (or near equal) amounts of air and fuel, the quality of the mixture in each cylinder can vary—creating unequal cylinder-to-cylinder torque output. Reading the tops of the pistons is a quick method of determining whether you have different mixture properties in different cylinders. Adjusting individual cylinder spark timing helps, but unless or until you are able to equalize the burn process among the cylinders, you're still dealing with multiple engines of unequal and unoptimized power. For more on the subject, see the next chapter.

HOW CAN YOU TELL IF IT'S WORKING?

It bears repeating that improved mixture quality tends to accelerate flame rate. As a result, it creates some conditions that do not require sophisticated measurement techniques. By producing an air/fuel blend with smaller droplets of a more uniform size, flame travel will be both smoother and quicker. This conditioning also includes the reduction of puddling, which results from mechanical separation of air and fuel.

Here is an example: If flame speed is increased, less initial spark timing is required to complete the combustion process. The later the spark, the less pressure will be exerted on the piston prior to TDC on the compression stroke. So, if you are working on the objective of improved mixture quality and discover less spark timing is required for optimum on-track performance, you're on the right path.

There's also a chance you'll discover an improvement in fuel economy. Even if mandatory pit stops aren't on the agenda, saving some money on the cost of current fuels can help—certainly over the length of a racing season.

Also, see if you can take some jet out of the engine or simply reduce net fuel flow. Generally, if mixture homogeneity is improved, less fuel is required to make the same (or more) power as before. Again, this can represent a fuel savings. At the same time, throttle response and off-the-corner torque tends to be sharper. You will find the engine accelerates through the rpm range more effortlessly. Stated another way, "transient torque" improves.

If you're measuring exhaust gas temperatures, faster burn rates usually translate into lower exhaust gas temperatures. The net effect of this is that more heat (work) is released within the combustion space. Therefore, temperature in the cylinder will be lower at the time of exhaust valve opening.

IN SUMMARY

Mixture (or air) motion in the combustion space is affected by several variables. Racing rules don't always permit modifying or using parts to improve this motion. In the best case, you'd like to begin flow-quality improvement early in the inlet path and continue it all the way through the combustion process.

However, if you're allowed to make changes to the piston crown configuration, specific benefits can be derived. Piston-head shapes that improve flame speed are also a good way to maintain—and sometimes even improve—available horsepower, even as sanctioning bodies mandate ever lower compression ratios. But even if rules prevent you from making any changes at all, the ability to read the combustion chamber floor is a positive step toward finding ways to optimize engine performance.

DOES IT REALLY WORK?

You be the judge. Following are data gathered from the testing of two engines: one at a maximum torque output of 363 lb-ft and a second at 400. In each instance, the engines were built to the same mechanical specifications (camshaft, static compression ratio, and displacement). For best power, we learned that the spark-ignition timing needed to be retarded slightly from the baseline timing, further indicating an increased rate of burn and an opportunity to reduce negative torque. The only difference between the two engines was the replacement of base configuration pistons with Hi-Tech's Swirl Quench pistons, which are designed to promote fast-burn characteristics. All torque data are corrected for environmental conditions (CBT stands for Corrected Brake Torque).

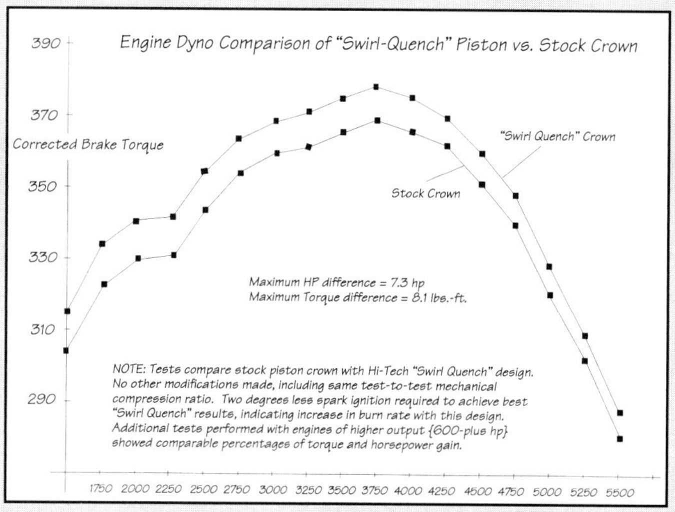

As is the case with most modifications that enhance burn rate, the time available to improve combustion rate is a function of rpm. The higher the rpm, the less time available to speed up the burn. Accordingly, note that gains with the Swirl-Quench concept tend to decrease slightly at higher engine speeds. However, along with changes to valve timing that are consistent with faster combustion, these losses can become gains comparable to lower rpm improvements.

Test Engine Package No. 1 (200-rpm/sec Acceleration Test)

RPM	Baseline CBT	Swirl-Quench Pistons CBT
3,500	386	400
3,600	377	392
3,700	379	393
3,800	377	392
3,900	377	392
4,000	381	393
4,100	382	391
4,200	385	391
4,300	383	391
4,400	383	391
4,500	383	391
4,600	382	389
4,700	380	386
4,800	379	383
4,900	376	386
5,000	374	380
5,100	374	376
5,200	368	374
5,300	361	375
5,400	356	368
5,500	348	361

Test Engine Package No. 2 (250-rpm Step Test)

RPM	Baseline CBT	Swirl-Quench Pistons CBT
1,500	301	314
1,750	326	334
2,000	329	338
2,250	330	330
2,500	345	352
2,750	356	364
3,000	363	370
3,250	365	371
3,500	368	375
3,750	371	378
4,000	366	374
4,250	361	367
4,500	348	355
4,750	333	340
5,000	319	327
5,250	299	307
5,500	277	285

Note: It is important to remember that the modification of a piston crown only addresses a portion of the combustion space. Given an opportunity to configure the cylinder head (the combustion chamber roof) to complement the piston crown (or floor), it's possible to achieve greater increases in power. However, if rules or parts availability do not provide an opportunity to adjust combustion-chamber shape or surface texture, piston crowns that aid mixture motion offer an opportunity to enhance combustion efficiency, on-track fuel economy, throttle response, and net power.

Chapter 10

READING COMBUSTION PATTERNS

Combustion Surfaces Can Tell You Much About An Engine's Performance, or Lack Thereof

Text and Illustrations by Jim McFarland

Although showing some signs of air/fuel mixture variations across this piston's top, here is an example of relatively good crown coverage and a lack of mechanically separated air and fuel. Note the intake valve relief shows color around the crown's deck surface. Other photos will reveal how air and fuel may separate in this area.

It's likely you've looked at the combustion patterns left on piston tops and cylinder head chambers. Here's a "Saturday Nighter's" guide to understanding what you see and how these patterns can be related to steps for increasing power and parts life. Study the photos because they're the keys to finding hidden or lost engine performance.

Simply stated, we'd like air/fuel mixtures in an engine's combustion space to be uniform throughout, both prior to and during the "burn." If this were possible, the combustion process would move at a relatively uniform rate from beginning to end while avoiding detonation and optimizing power. But it's not. There are numerous and complicating factors that tend to prevent this ideal situation. However, by working to produce uniformity of air/fuel mixtures, chances are we'll boost combustion efficiency and gain power. The objective of so-called "mixture motion" (swirl, tumble, and so on) is to improve the homogeneity of fuel and air before and during combustion. The goal of this story is to provide some insight about what post-combustion "footprints" in the combustion space can reveal.

CONDITIONS THAT RELATE TO POOR COMBUSTION EFFICIENCY

Let's say you have an engine that requires an excessive amount of ignition spark timing to run correctly. That covers numerous causes, but consider the following as it relates to combustion efficiency. If the combustion space is contaminated with residual exhaust gas, chances are additional timing is required to extend the burn time with the intention of building as much chamber heat as possible. All else being equal, heat is power. Exhaust gas contamination of fresh air/fuel charges reduces heat, and a racing engine with built-in "EGR" isn't going to fulfill its power potential.

Suppose you've discovered an inordinate amount of fuel is required to optimize power. The possibility exists that excessive mechanical separation of air and fuel is the cause, requiring additional fuel to compensate for unburned and separated fuel passing directly out of the exhaust ports. Such cases often require additional spark timing, too—both conditions are undesirable. Remember, the later you can initiate spark, the less "negative" torque will be produced on the crankshaft. Even though increased spark can improve power, it's at the expense of work being done against the crank's normal direction of rotation. Poor mixture quality, including air/fuel separation, can lead to excessive spark timing requirements.

Improper matching of piston crowns and combustion chambers can also lead to mechanical separation of air and fuel. Unless good levels of mixing (homogeneity) are achieved, portions of induced mixtures will be disrupted (separated from the air) and reduce net power. Increased mechanical compression ratio achieved by allowing incorrectly shaped piston crowns to invade combustion chambers is a common cause for such problems. What you may gain by increasing the mechanical compression ratio can be partially lost by reduced mixture quality and decreased combustion efficiency.

If there's a rule of thumb embedded in this discussion, it points to the need for staying focused on the combustion space as you look for ways to make power or the reason a given engine package isn't living up to expectations. This area may become a window to an engine's ability to efficiently convert fuel and air into heat or power. Developing the skill to read combustion patterns is time well spent, especially if you intend to assemble parts packages that make good power.

While the quench area (typically clean in higher compression engines) and crown near the spark plug are acceptably absent of color, note areas in the valve pockets where air and fuel have separated. This tends to cause overly rich conditions further in the burn area as noted across the piston dome.

CLEAN AREAS IN THE COMBUSTION SPACE

Are these acceptable? Not necessarily. Areas not exhibiting combustion residue frequently suggest fuel "wash" or the possibility of separated air and fuel passing over and causing the clean surface. For example, with its intake valve removed, look down an intake port toward the combustion chamber.

A line-of-sight path often indicates the path of fuel flow, which is not always the same path as air. The relative incompressibility of fuel (compared to air) and the greater-than-air level of kinetic energy it develops moving in excess of 300 feet/second during a typical intake cycle (around peak torque and above) can lead to air/fuel separation during changes in flow path direction. At higher rpm, flow rate increases accordingly. Therefore, as you peer into the port, the back side of the combustion chamber may exhibit a clean area of fuel wash. Such areas are candidates for the so-called "dimpling" process, a method in which air conditioning is used to improve fuel conditioning.

In a similar fashion, clean areas on piston crowns can be further indications of separated air and fuel. That's the worst case. In the best case, these conditions suggest abrupt changes in air/fuel ratios caused by interruptions in the movement of homogenized air and fuel moving through the combustion space. (By examining the accompanying photos, you'll see visual examples of descriptions being provided in the text.) Ideally, the smaller the combustion chamber and the flatter the piston crown, the higher the probability of reducing mechanical separation of air and fuel as mixtures pass through the combustion space, at least as far as piston-crown intrusion is concerned. What follows is an opportunity to increase flame rate, thereby reducing initial spark timing requirements and minimizing negative torque on the crank. Combustion processes, disrupted by wide ranges in air/fuel ratio caused from poor mixture quality, are at the heart of optimizing engine power. Remember, the combustion spaces can be a window to increasing engine output. Don't read 'em and weep. Read 'em and do something about what you see.

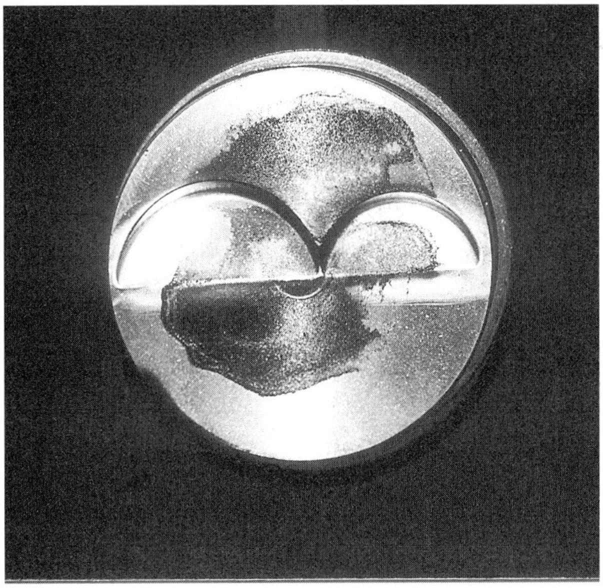

Here you can see a more advanced case of air/fuel separation around the valve pockets and an amount of fuel "wash" across the plug side of the crown. Both heat and working cylinder pressure are being lost under these conditions. A disproportionate amount of fuel is usually required in an attempt to compensate for the loss in power, although this practice is largely a compromise that doesn't correct the problem.

Does the color of combustion residue mean anything? It can. The type of fuel being used has a measure of influence. Alcohol tends to burn cleaner and leave lighter-colored residue, compared to gasoline. In either case, you'd like to avoid a wide spread in color, light to dark. Darker areas suggest

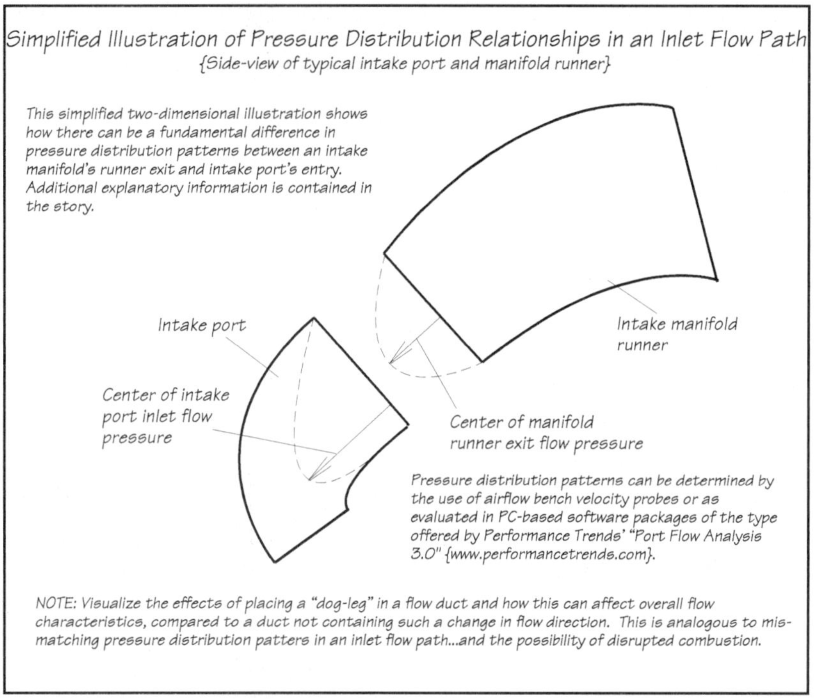

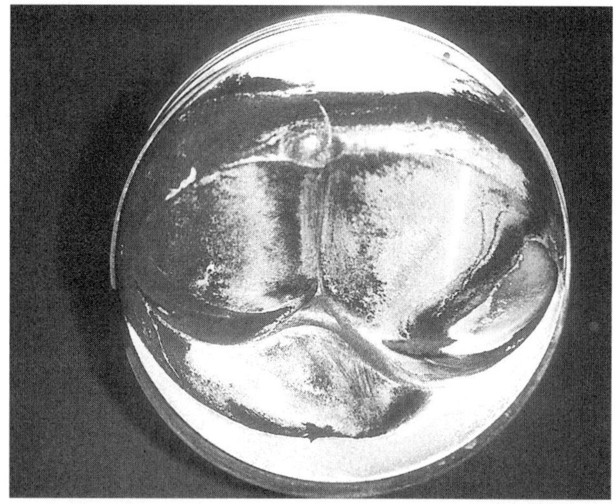

Here you can see the evidence of a fairly wide range of a/f ratios in the combustion space. Note mixture richness along the back side of the dome (around the spark plug area) and a worse condition as the burn progresses to the exhaust side. This photo suggests excessively rich mixtures trailing around the piston's perimeter, resulting in an inefficient burn and lost power.

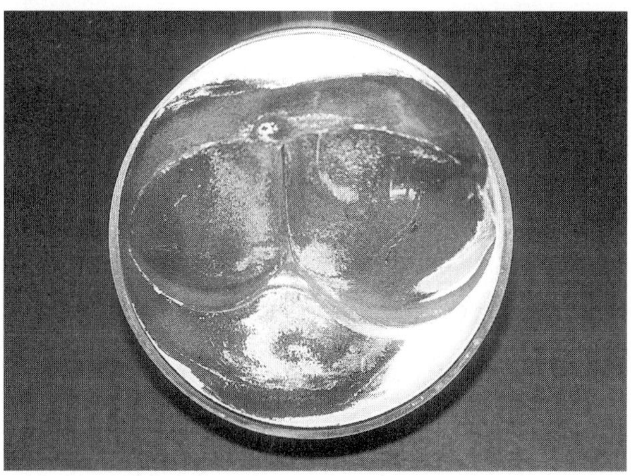

In this photo, primary mixture motion was counter-clockwise. The pattern indicates separation of air and fuel in the intake pocket and rich conditions against the combustion chamber's backside wall on both the intake and exhaust sides. Note how rich the burn becomes at about "3 o'clock" and how it turns leaner as it progresses onto the crown's deck. Not bad, but another case of mixture variation throughout the combustion space.

the combustion of rich air/fuel mixtures while lighter ones point to lean ratios. Flame rate also slows as mixtures become increasingly fuel rich, again suggesting the need for earlier spark timing (to be avoided where possible).

If you spot areas much darker than other locations, particularly on piston crowns, it's reasonable to assume excessive fuel may have puddled and been lost before combustion was complete. This is evidence of rich, slow-burning mixtures. In such cases, look for very light-colored areas nearby. These may be indications of very lean mixture conditions and suggests detonation or near-detonation conditions. If you suspect this problem exists, even short of mechanical damage to piston rings or ring lands, inspect the rod/crank bearings for further evidence. An engine can run for a period of time in threshold detonation, lose power, and not damage piston assemblies, which can be detected by reading piston-crown combustion patterns and inspecting "hammered" bearings.

Especially in engines using flat or near-flat piston crowns, you can expect most of these areas to be pretty much covered by combustion residue, hopefully in a shade of gray without significant variation in color contrast. Such engines can have either low mechanical compression ratio or have relatively small combustion chamber volumes and higher compression. In either case, if the pistons are notched for valve clearance, watch for clean crown surface areas immediately past sharp edges accompanying the notches. To help prevent mechanical separation of air and fuel in these areas, a gentle blending of the edges tends to help. Increased burn efficiency and power often replace what you may lose in compression ratio.

Can conditions upstream of the combustion space affect the burn? Certainly. In particular, the distribution of pressure between an intake manifold's runner exits and cylinder head intake ports can be especially influential. To improve an understanding of how this works, take a moment to study the included illustration showing a two-dimensional relationship between a runner's exit flow pattern and an intake port's inlet flow pattern. You will see two different pressure distribution patterns in the illustration. Because of these differences, there can be an exchange of energy at the junction between manifold and head, causing a

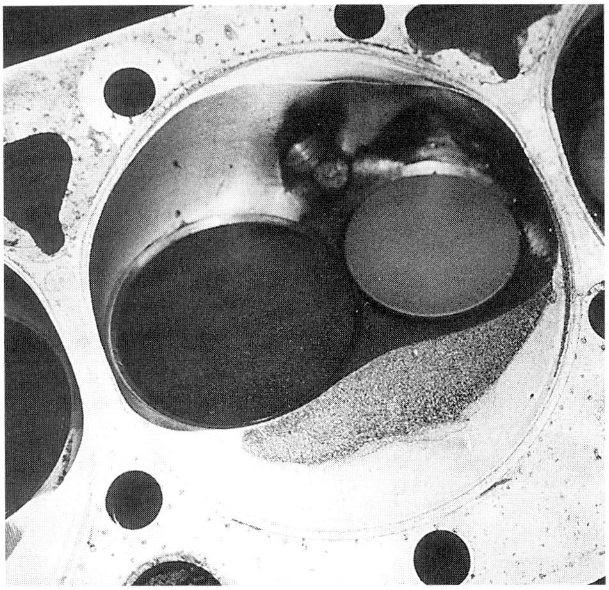

Typical of both gasoline and alcohol-fueled engines, fuel wash can occur on a line-of-sight path from the intake port to the combustion chamber back wall. While this photo is of a small-block Chevrolet V-8 chamber, similar conditions can exist in other brands/designs of cylinder heads. Note the clean area adjacent to the spark plug and evidence of an excessively rich burn just past the plug where separated fuel re-collects into a rich condition. Dimpling of clean surfaces helps restore boundary layer activity and improve burn efficiency.

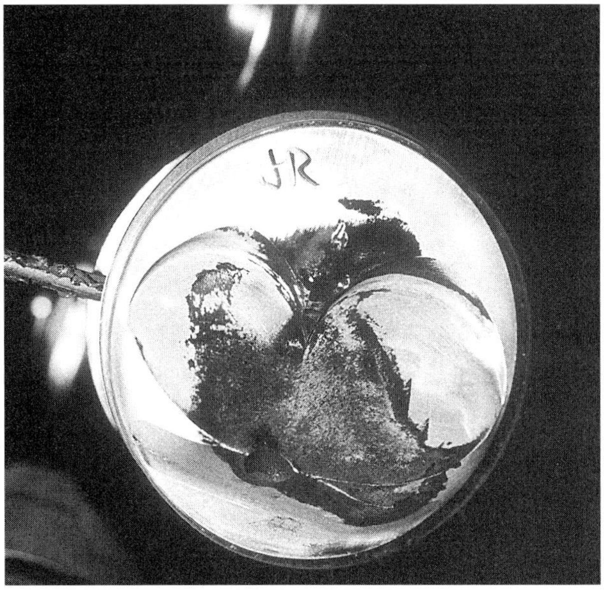

This engine was in trouble. Fuel was washing around the chamber and collecting toward the center portion of the chamber. Incidentally, it was too rich there. Engines in this condition often require excessive amounts of spark timing and/or fuel to optimize power. Either or both of these requirements suggest a problem like this exists before heads are pulled and surfaces examined.

disturbance in the incoming air. While not always detrimental to mixture quality, it's possible this condition can cause downstream mixture motion, which is disruptive to an efficient burn. The goal is to match the profiles.

Furthermore, since air is compressible (certainly when compared to fuel), it can assume energy or directional changes more readily than the heavier fuel. If air changes direction quickly enough, a mechanical separation of it and fuel occurs, resulting in a possible alternation of air/fuel ratio downstream of the junction. Typically, there is adequate mixture motion in the combustion space as flow enters this area and is prepared for closing of the intake valve. Efforts to create desirable amounts of swirl or tumble are part of the intentional components of beneficial mixture motion. Other types of "motion," including those sometimes created at the manifold/head interface, may be harmful to supporting combustion efficiency.

To minimize this problem, the measurement and matching of manifold exit pressure distribution patterns with the intake ports is a worthwhile step to improving downstream mixture quality. This can be done on an airflow bench using velocity probes or by the use of the pressure maps capability often included in PC-based software geared to computer-aided airflow measurement. A typical example is Performance Trends' Port Flow Analyzer 3.0, which you can review on the Web at www.performancetrends.com.

It is important to understand that air is the medium by which fuel is communicated to the combustion space. Therefore, air quality must be considered in order to communicate fuel in its most combustible form. This is especially true when certain techniques are used to intentionally create specific mixture motion patterns, which can be disrupted by peripheral conditions (like the manifold/head interface issue just described).

CONCLUSIONS FOR THE "SATURDAY NIGHTER"

Of the benefits derived from learning to read combustion patterns, the potential for power gains through improved combustion efficiency is principal. Clean areas just past sharp edges in the flame path typically cause air/fuel separation and wasted fuel that could otherwise be used for power. It's important to realize that modifications to combustion surfaces (pistons in particular) that cause slight decreases in mechanical compression ratio can lead to increases in power from improved burn efficiency.

This writer has seen numerous cases where changes to piston crowns that reduced compression ratio by up to a half-point still produced 3 to 5 percent net gains in power with the same volume of consumed fuel and less spark timing. If it becomes necessary to create increases in compression ratio at the expense of sharp-edged disruptions in the flame path, perhaps the higher compression ratio is of little value. Remember, combustion is not an

Here's an advanced stage of air/fuel separation. Note the fuel-washed areas on the chamber's back wall near the intake and exhaust valves. An excessively rich condition existed in the channel beginning at the spark plug and progressing across the chamber to the quench area. Cylinder pressure conditions, under such adverse circumstances, reduce uniformity of flame travel and pressure rise, causing a net reduction in power. Mismatches between intake manifold runner exit and intake port entry velocity patterns can lead to this version of disrupted combustion efficiency. Depending upon an engine's horsepower potential, these problems can reduce power in the range of 4 to 8 percent per cylinder.

event—it's a process. Visualized in very slow motion, consider it much like setting a sheet of paper on fire. Once begun, the flame moves progressively throughout the combustion space (although much quicker), similar to the burning of the paper.

Therefore, in reading about or modifying piston crowns and combustion chambers, identify and provide for smooth flame movement. Sharp-edged valve reliefs or piston crowns can contribute to mechanical separation of air/fuel mixture and reduced combustion efficiency, as noted in the accompanying photographs. When a problem arises, it's usually characterized by very dark combustion residue (if an overly rich condition exists) or clean areas that often signal fuel wash and the loss of burn efficiency.

A significant amount of data suggests a large percentage of liquid fuel (or improperly conditioned fuel) becomes atomized in the combustion space after the intake valve closes. This is the result of heat and pressure during the compression stroke. Even though this condition appears verifiable, it's still a wise decision to work toward inlet air-flow structuring that enhances fuel atomization and suspension along the entire induction path, prior to the inlet valve closing, regardless of where fuel is introduced into the air stream. The more you can condition pre-compression stroke mixture or air quality, the less reliance is placed upon what happens after the intake valve closes.

Finally, there's another area in the combustion space that is often a source of lost power. In the interest of gaining airflow, chamber back walls are sometimes laid back to improve line-of-sight flow. This is frequently the result of cylinder head modifications made in alcohol-fuel engines where increased air is required for the reduced latent heat of combustion these fuels provide compared to gasoline. Once again, a trade-off may be required to optimize net airflow but not at the expense of poor flame travel. Observing combustion patterns is a key to solving this problem.

Reading combustion chambers and piston crowns is a "homemade" approach to gauging the results of the airflow/combustion compromise, allowing additional changes to be made in the combustion space after an engine has been run and has produced residue patterns. In these cases, the old axiom, "More flow isn't always more power," might be modified to read, "More flow isn't necessarily more power unless air quality supports good combustion efficiency."

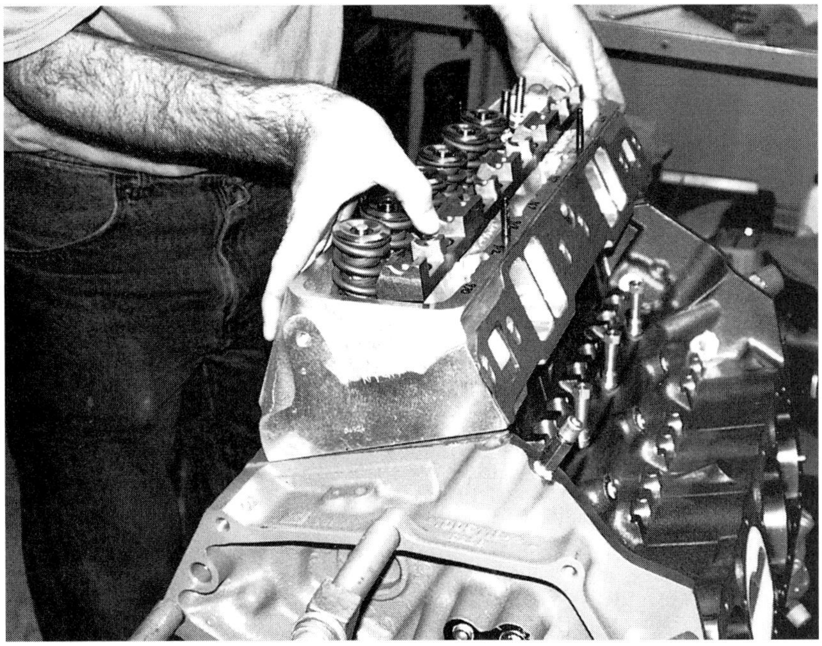

Chapter 11

THE BASICS OF GETTING A HEAD

Fundamental Cylinder Head Features, Functions and Modifications

Text and Illustrations by Jim McFarland

Successfully integrated engine packages can be a matter of luck. But this is neither the most intelligent nor the best way they should be assembled. Aside from material type (aluminum or cast iron), cylinder head choice is most often made on the basis of airflow numbers or port volumes. In fact, both factors can influence incorrect overall decisions. The most important factor is how efficiently the head helps convert air and fuel into power. It is in the combustion space (cylinder) that this begins, and cylinder heads (including their port-flow capabilities) are key to making power. Let's look inside the combustion space to see how different cylinder heads can affect how well the fuel mixture burns.

INSIDE THE COMBUSTION SPACE AND THE EFFECT ON POWER

The combustion roof and floor: If we compare the combustion space to a room, the walls are the cylinder, the floor is the piston top, and the roof is the combustion chamber of the cylinder head. The relationship of the roof to the floor is critical since each approaches the other prior to, during, and after combustion. This relative motion between the roof and the floor affects the quality and quantity of air/fuel charges and exhaust gas as they enter, reside in, and leave the combustion space. Cylinder heads, therefore, can have an important influence on how mixtures are processed to make power.

If domed pistons are used, the shape relationship between the combustion roof and the piston top is critical to maintaining then initiating proper flame travel. Here's why: If you consider the combustion process similar to burning a flat sheet of paper, the flame progresses from burned to unburned parts of the paper. At a much faster rate, the combustion flame initiates near the spark plug and traverses the combustion space—while the combustion floor approaches the combustion roof. Turbulence, created by the flame combined with movement of the mixture left over from its flow into the cylinder, moves across both the floor and roof surfaces. Any protrusion from the floor (piston) that interrupts flame travel can alter mixture homogeneity. This causes changes in air/fuel ratio and power—usually for the worst.

It is no secret that a specific relationship between the combustion roof and floor can benefit power. For example, if the relative shapes of a combustion-space floor and roof tend to "drive" compressed material (usually during the burn) toward the exhaust side of the space, a shorter distance must be traveled during the exhaust cycle. This tends to hasten flame rate and increase the potential for an improved exhaust cycle, both of which can aid power.

Mixture Motion and Port Design

Specific air or mixture motion can be established by designing specific characteristics into the intake path (either the intake port, manifold, or both). If this motion tends to be rotation in a horizontal plane (typically around the cylinder-bore axis), it is called swirl. If motion is in a plane perpendicular to the bore axis (somewhat like tumbling down a flight of stairs), it is called tumble. Both swirl and tumble can be generated in the inlet ports of a cylinder head, and several contemporary cylinder head providers (including OEM) use these techniques.

Even though the generation of swirl and tumble often reduce net air flow, the ability to increase homogeneity in the air/fuel charges can increase power and on-track fuel economy. Here is the point of emphasis regarding swirl and tumble: Both offer potential losses in net airflow, but both are tools for increasing power because airflow quality is equally important as quantity.

Air Flow Research (another provider of five-axis CNC porting services) utilizes combustion techniques evolving from its three decades of port-and-chamber study. An early user of the kidney-shaped chamber, the cusp (opposite spark plug hole) further enhances mixture motion (swirl) and movement of charge direction toward critical exhaust valve location.

Inlet and Outlet Flow

It is important to understand that air (or air/fuel mixtures) entering a combustion space is influenced by the geometry of an intake port, valve seat dimensions, valve size and shape, and combustion-chamber design. Even though each of these areas can affect the volume of "working fluid" passing into the engine, the quality of this substance must also be considered. Unless air and fuel are properly mixed, combustion efficiency will suffer. If you are going to build a competitive engine, you must consider both airflow quantity and quality. Try burning air without adequately blended fuel. In the words of one widely acclaimed cylinder head modifier, "You can't tune a summer breeze." For example, modifying the "back wall" of a combustion chamber just to achieve more airflow may be disruptive to proper mixture motion before and during the burn. As a result, you get the increased airflow you were looking for but at the cost of reduced power.

A similar set of conditions exists on the exhaust side of the combustion space. Exiting exhaust gas, initially under high pressure, needs to lead to a high level of cylinder evacuation. This is important since residual combustion material won't burn a second time; you don't want unburnable residue hanging out in the combustion space.

During the expansion stroke, the timing for opening the exhaust valve is critical to maintaining work on the piston. A trade-off exists between maintaining useful pressure (work) on the piston and opening the exhaust valve early enough to optimize blowdown pressure for cylinder evacuation. Low-lift exhaust-port flow capability is important during this time. The key is to utilize the residual cylinder pressure to help start and maintain exhaust flow until piston motion begins displacing

This recently introduced piston from Hi-Tech Engine Components utilizes a unique top shape (the combustion-space floor) to direct mixture motion as swirl; it increases air/fuel mixture homogeneity by the addition of strategically located "dimples" on the surface.

the remaining material. When you are evaluating different cylinder heads, don't forget to consider low-lift exhaust port flow characteristics, which are also influenced by valve and valve seat configurations.

Intake and Exhaust Port Sizing

Take a deep breath; what you are about to read may not be what you expected. First, let's examine the issue of port volumes. Granted, cylinder head ports (particularly intake ports) have been designated—even related to power levels—on the basis of cubic centimeters. Use your imagination and visualize how many different ways you might configure a 190cc intake port. You may have even seen correlations offered between port volume and specific piston displacements.

Perhaps there is a relationship between port flow and port volume, but it would seem secondary to a given port's potential for net flow and how this capability relates to volumetric efficiency—which is related to power (or torque). So rather than assume specific port volumes and their possible relationship to engine size, it appears to be more prudent to include port-flow numbers (although these are based on steady-state flow conditions) and the volumetric efficiency potential (and requirements) of a particular engine package.

At the risk of oversimplification, let's run some numbers. Assume we have a 350-cid engine that's turning 6,000 rpm. We know that half of this displacement is addressed every crankshaft revolution, so we need to supply air to 175 cid per crank rotation. With a little arithmetic that includes converting cubic inches into cubic feet, and assuming 100 percent volumetric efficiency at this rpm, we need 608 cfm of air per revolution, or 152 cfm per cylinder per crank revolution.

Now, wouldn't it be easier to relate these approximated numbers to a steady-state airflow bench where you're frequently (but maybe not always) examining the flow characteristics of a single port, which in turn is feeding into a single cylinder? It may be that a specific port that provides suitable air to a cylinder of this size (at this rpm) actually measures a certain volume, but isn't this a quantity determined after you've optimized the port's design? Can you then hypothesize that a port of this volume is what is required for a cylinder of this displacement, or does this previously worked-out port coincidentally contain a certain volume—as will any port? It's worth a little thought.

SIMPLIFIED RELATIONSHIP OF MEAN FLOW VELOCITY & TORQUE OUTPUT

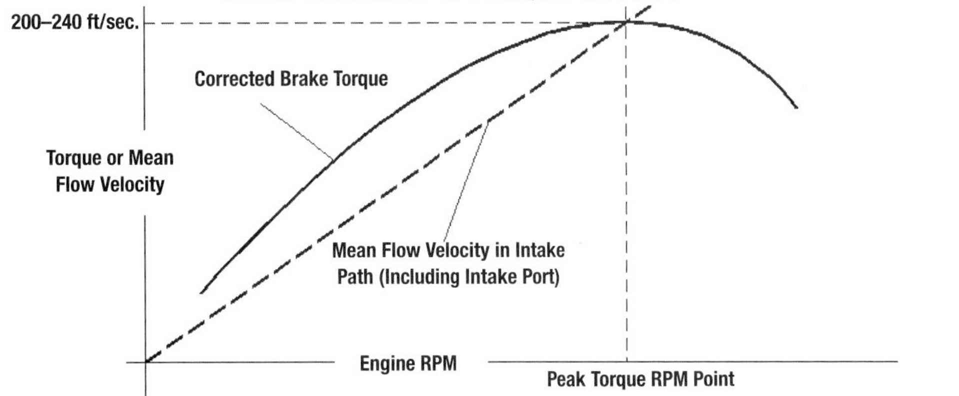

In this illustration, some assumptions are required. Mean flow velocity is taken to be a value bounded by the extremes present in the mist path (zero to maximum). It is also the value at which the "working fluid" (air, fuel or air and fuel) is moving most of the time. A value of 240–260 ft/sec. is commonly accepted. In fact, overall flow rate will begin to diminish near peak rpm, the result of pumping losses and a lack of time to produce volumetric efficiency. Port flow velocity is a combined function of piston displacement, rpm and cross section area of the passage. Therefore, decisions on port sizing should include where an engine is expected to be operated most of the time, relative to the production of torque.

ILLUSTRATION SHOWING PRESSURE DISTRIBUTION SHIFTS VS. FLOW PATH GEOMETRY CHANGES

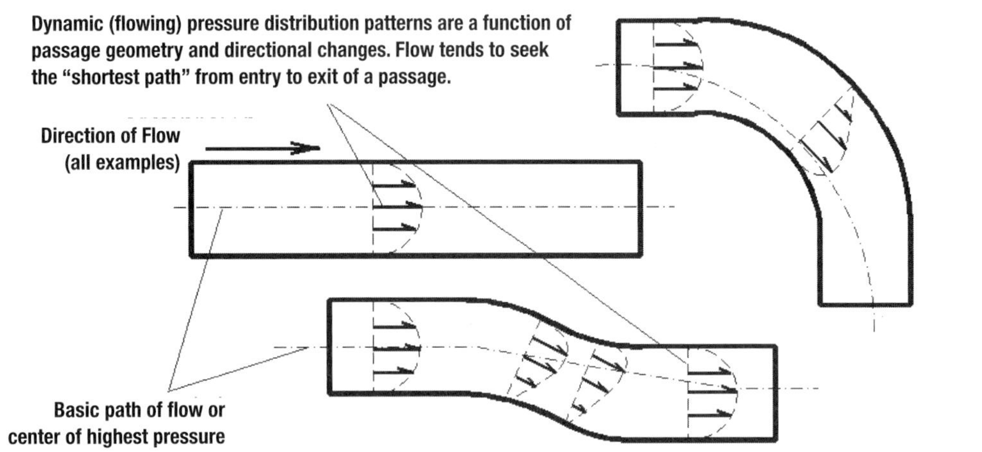

Whether the "working fluid" is air or air and fuel, air tends to follow the shortest geometrical path between two points of pressure difference. In areas where pressure distribution is "skewed" to one side of the path to another, low pressure opposite these areas tends to encourage air/fuel separation or low pressure air instability. Intake and exhaust port dimensioning (or modification) should include consideration of these basic features, attempting to keep pressure distribution as uniform as possible across each port section.

HEAD MODIFICATIONS

In the predictable fashion of merging technologies, the advancement of both the cut-and-try method of cylinder head modification (otherwise known as learning by trial and error) and computer-based programs that can accurately model air/fuel mixture flow in an engine is clearly affecting the grassroots circle-tracker. The trickle-down effect of information carefully guarded by leading Nextel Cup teams and OEM research labs is finding its way into parts used by the amateur racer. There seem to be two main reasons for this. One is that precision modification of heads on CNC machines is becoming more affordable. The other is the advanced science of combustion theory is moving out of the confines of the scientific community and into race shops, where it is getting put to the test. It is this advanced scientific understanding of what goes on in the combustion space that is fundamental to the conclusion of our discussion on cylinder heads.

Port Sizes, Shapes, and Modifications

We must first recognize that intake and exhaust ports have a particular effect on how the "working fluid" (air and fuel, or exhaust gas) is directed into and out of the combustion space. Both types of ports are also situated in the bowl region where a directional change occurs. This makes their short-side (the path of shortest distance for the air/fuel mixture to reach the combustion space) areas especially sensitive, particularly during the closing

ILLUSTRATION COMPARING BASIC PORT SECTIONS FAVORING SHORT-SIDE FLOW PATHS

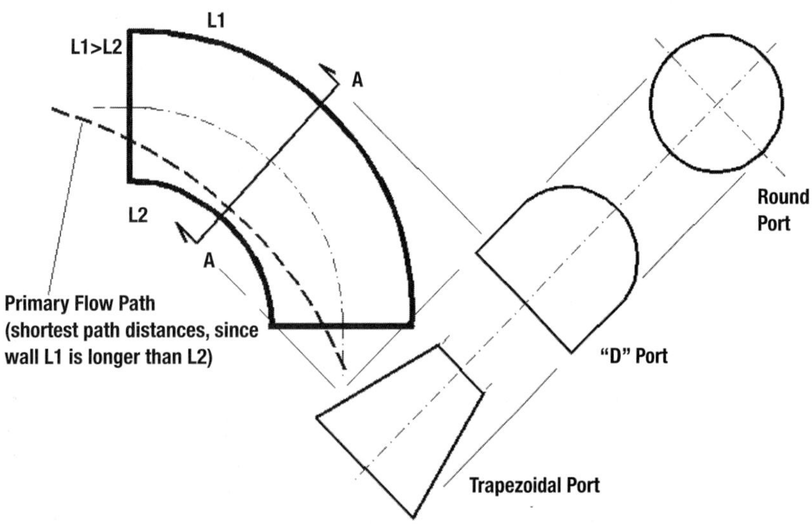

Given two dissimilar path wall lengths (L1 & L2), flow enhancement can be produced by treatment of the short-side floor. "D" port improves flow, trapezoidal shape further aids flow by increasing and decreasing surface areas (top to bottom) and making pressure distribution across trapezoidal section more uniform.

BASIC FLOW PATH CHANGES IN INTAKE AND EXHAUST PORTS AS A FUNCTION OF FLOW RATE OR RPM INCREASE

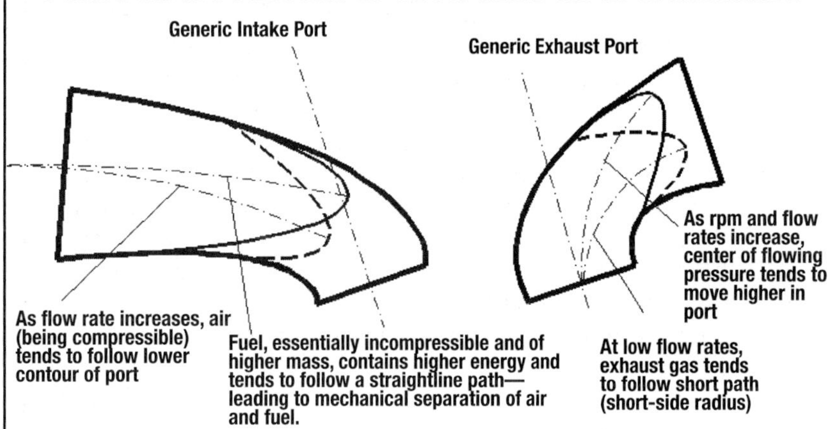

When selecting or modifying intake and exhaust ports, it is important to consider the changes that occur in flow "quality" and location in ports, as flow rates or engine rpm increases. Gains, or losses, in net flow can develop by subtle changes (or designs) in intake and exhaust ports.

portions of an intake or exhaust cycle. The shape of intake and exhaust ports (in these short-side areas) is made more critical because of the directional change mentioned.

We discussed port sizing earlier in this series. Aside from relating port volume to piston displacement, there is another point to consider. We know that flow velocity is not constant, but there is also a mean flow velocity (m.f.v.) associated with working fluid movement into or out of an engine. Put simply, an m.f.v. value exists at any rpm and is primarily dependent upon passage cross-section area, rpm, and piston displacement. At peak torque (peak volumetric efficiency), the value of m.f.v. is generally accepted to be 240–260 feet/second (see illustration p. 65).

If a value of 240–260 feet/second can be expected at peak torque, then it is possible to select specific port (or passage) cross sections that generate this m.f.v. at a desired rpm. Stated another way, if we want to provide a torque boost at 4,500 rpm (based on a specific piston displacement), it's a matter of sizing the passage to produce this m.f.v. at 4,500. Intake (and exhaust) ports and passages respond to this method.

Here is a greatly simplified equation that will help approach the problem of port sizing:

Port Cross-Section Area =
(Cylinder Volume x rpm) ÷ 88,200

- Port cross-section area is in square inches
- Cylinder volume (one cylinder) is in cubic inches
- rpm is engine speed at desired torque boost
- 88,200 is a constant in the basic equation

By using a little algebra, any of the three variables (port cross-section area, cylinder volume, and rpm) can be solved if you know the other two. For example, if you know the size of the intake- or exhaust-port passage, including manifold or header passage, and want to determine the rpm at which it produces a torque boost, the equation becomes the following:

Torque Boost rpm = (Port Cross-Section Area) ÷ (Cylinder Volume x 88,200)

Correspondingly, if gearing fixes a particular rpm where torque is required, and no changes can be made to a particular set of ports (intake or exhaust system), engine sizing can be calculated for these predetermined values.

While this simplified approach to port (intake/exhaust passage) sizing may appear unrealistically simple, we invite you to check it against parts packages for which on-track (or dyno) data already exists. The results may add to your confidence in this mathematical approach to basic passage-size dimensioning.

There is also the issue of port taper vs. non-taper. Our objective is not to advocate either approach for intake or exhaust port dimensions. However, the following thoughts are offered for your

CONTEMPORARY PORTING TECHNIQUES

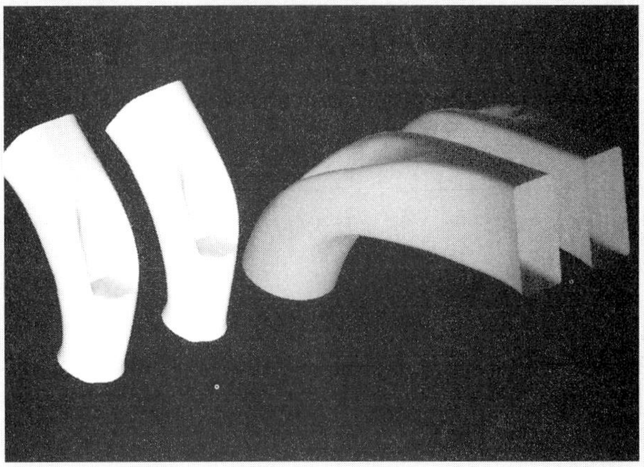

Solid models are determined on a computer screen. Each set of complete port dimensions is included for the purpose of tool-path generation. Any changes to port configuration are accompanied by a corresponding modification to the tool path required to produce the design change.

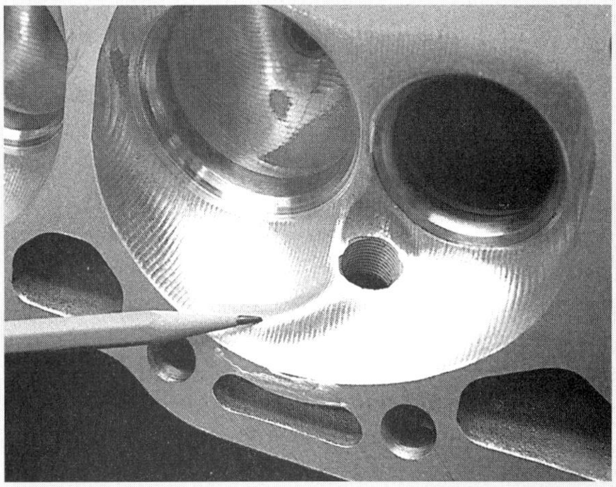

Even the smallest of details can be built into the CAMplete Port program, such as this "flow director" used by CNC Cylinder Heads to create motion in the form of swirl and tumble, both used to enhance air/fuel mixture quality and improve low- and mid-rpm torque output.

Since the early days of modifying engines for specific purposes, the modification of cylinder heads has been common practice. Early techniques largely centered on basic port enlargement and surface finish (polishing), founded on the "bigger is always better" philosophy. This was also well in advance of the development of airflow benches. When these benches emerged in the '60s, "experts" abounded, touting their ability to create increasingly higher levels of airflow by a variety of porting methods.

A sequence of events followed. Port-to-port airflow equality, or scatter, became important to the modifiers. Low-lift and reverse flow analysis followed. Next came an awareness of the differences between and the importance of airflow quantity and quality. By this time, enterprising porters enlisted the services of CNC operators and skilled digitizers to replicate port shapes, and the move was on to computerize the science of cylinder head modification. In the marrying of technologies, however, it is not uncommon to find certain discrepancies that prevent a cost-effective use of the new thinking. Such has been the case with CNC porting techniques.

CNC Cylinder Heads (www.cncheads.com) is about to introduce a software package labeled CAMplete Port. It addresses previous problems associated with five-axis CNC porting of cylinder heads—specifically, the conversion of scanned data describing port dimensions into the nurbs, or 3D computer images that represent port shapes. Scanned or digitized information also tends to dilute specific point-by-point dimensional locations, requiring further processing of the information, typically by a CAM utility program or comparable data-smoothing operation. According to CNC Cylinder Heads' Bob Hudgins, "More times than not, this type of utility alters the position of the data points and violates the original port design. CAMplete Port reads the data exactly as measured and rapidly produces very accurate surfaces."

The benefits of CNC porting techniques are numerous. Among them is the ability to repeat cylinder head porting combinations that produce predictable power levels, rebuild after rebuild, engine to engine, head to head, or even season to season. Consistency in producing internal port shapes, valve guide-boss contours, and combustion chambers also is significantly improved. Time-to-completion of porting jobs is accelerated as well. And with the current emphasis on building and tuning engines one cylinder at a time, CNC porting methods help ensure the consistency of port-and-chamber modifications that this method requires.

Small cylinder head shops, particularly those previously relying on outside CNC services for this type of work, can now obtain a software package that eliminates the need for highly skilled machine operators and/or extensive training. As a Windows-based software system, CAMplete Port may offer an expedient path to obtaining the benefits of CNC porting with a minimum of effort, compared to previous methods and expense.

consideration to determine which approach may be preferable.

If you accept the concept of m.f.v., then it should be relatively easy to understand that the port cross section affects the flow rate (in addition to rpm and cylinder volume). At any given rpm, increasing (or decreasing) the cross section will change the flow velocity. For example, if we would like to obtain our 240–260 feet/second flow velocity at peak torque rpm (based on a certain cylinder volume), a specific cross section will enable that to happen. If that area is increased, a higher rpm will be required to restore the 240–260–feet/second values. A decrease in area would lower the rpm at which this velocity would be achieved.

One school of thought is that when the optimum m.f.v. is reached, all working fluid in the passage should be "moving" at this velocity to maximize volumetric efficiency (total contents in the cylinder at the time of intake valve closing). If the passage has taper (cross sections larger and/or smaller than the area producing the 240–260–feet/second rate), the volume of fluid moving at this rate will be minimized (see illustrations p. 65). Without taper, a greater volume will be moving at 240–260 feet/second when this rate is achieved, thereby optimizing the cylinder's contents at intake valve closing.

A different concept promotes the value in having taper in an intake or exhaust passage—especially intake. Utilizing a combination of compression and expansion waves from passage entry to exit, the wave-motion theory often involves a computational approach dealing with multiple-pulse analysis. It includes the relationships among intake and exhaust pulses in conjunction with the timing of intake- and exhaust-valve events. One commercially available PC software package that utilizes this method is Dynomation from VP Engineering (www.proracingsim.com). In a fashion similar to Engine Cycle Analysis, this program allows the manipulation and study of variables that affect intake, exhaust, and valve-timing processes, including port dimensioning. In a sense, it's like ECA on your PC, and you don't get dirty or spend money evaluating different parts or combinations. It's also an excellent educational tool.

Of course, the counter approach to computer modeling is the old cut-and-try method—sizing your ports and evaluating them on a flow bench. While this time-honored approach still has value, exercise caution to avoid the "more flow is always more power" concept, because in many instances, it's not. When flow benches appeared on the motorsports scene in the mid-'60s, modifications to stock parts were the norm. Today, there is a proliferation of stock parts that approach optimum performance right out of the box.

The use of a flow bench is often reduced to tailoring parts to meet the demands of a specific engine package or application. Velocity and boundary-layer probes (among others) are used to evaluate specific port designs, which can be consistently reproduced by using multi-axis CNC techniques. Still, in the hands of a skilled operator, flow benches can be valuable tools for engine builders and in cylinder-head-design shops.

WET FLOW AND PORT SHAPES

Believe it or not, these are related. Earlier, we discussed some examples of port shapes and their fundamental flow characteristics. What we discovered was the importance of maintaining equal (or nearly equal) pressure distribution patterns during directional changes in flow. While this is important for handling dry flow (no fuel conveyed), it is especially critical when the mixture includes fuel. Because air and fuel have significantly different weights, and fuel is essentially uncompressible compared to air, the two will separate when they experience a change in direction, pressure, or both. As a direct result, shaping ports to help establish and maintain uniform pressure distribution throughout all cross sections tends to aid fuel suspension. Even in dry-flow systems (since air behaves as a fluid), uniformity of pressure distribution helps ensure optimum physical dimensions of a port.

FAST-BURN BENEFITS

Discoveries dating back to the work of Sir Harry Ricardo and Michael May established the benefits of air/fuel charge turbulence in the combustion process. Ricardo noted (in the 1941 third edition of his book, *The High-Speed Internal Combustion Engine*), "The maintenance of turbulence until the time of ignition is probably the most important consideration of all, for upon turbulence depends the rate at which combustion takes place." Turbulence in the combustion space not only aids fuel atomization but also enhances flame speed and the rapid release of work—producing pressure on the piston.

Indicated Mean Effective Pressure (IMEP) is the difference between the pressure applied to the piston before TDC (compression stroke) and the pressure exerted after TDC (power stroke). Combustion spaces (the area between the head and piston surfaces) that promote a rapid flame tend to reduce the amount of ignition-spark timing (advance) required to complete the burn. As ignition timing is increased, so is the amount of "negative" (BTDC) work done on the piston. Conversely, the less timing required, the more net (IMEP) force is applied during the power stroke. Simply increasing burn rate, without any additional fuel or air, can increase net torque by the reduction in required spark timing.

The selection of or modification to a combustion-chamber design should be made on the merits of rapid flame travel, commonly noted by a lack of required spark timing. Engines that make the best power with minimum spark timing are typically characterized by a fast-burn combustion space. Recall that this includes effects of the piston top. Reduced exhaust gas temperatures are also a feature of fast combustion, since an increased amount of heat is confined within the cylinder for the production of power, thereby reducing exhaust gas temperature when the exhaust valve opens.

HEAD SCIENCE WITH LARRY WIDMER

Larry Widmer is a story unto himself. Emerging in the '70s with what he termed the "Soft Head," his advocacy of mixture motion and lean, fast-burn combustion has stood the test of time. A student of Sir Harry Ricardo and Michael May, Widmer brought fresh—if somewhat controversial—power-making concepts into racing, including NASCAR, during his tenure at Endyn (Dallas). Widmer was willing to share his wisdom about elements of fast-burn combustion, its benefits, and the future of circle-track engines for Saturday-night racers.

CT: Where do you consider mixture motion to be the most critical for the best power output?

LW: Assuming mixture (air/fuel) atomization is reasonable, engine efficiency (including the production of power) hinges on the quality of combustion. Since the cylinder is the final place of charge residence, it's imperative that any motion, prior to and during combustion, not degrade mixture preparation. The transition area from the intake port to the cylinder is responsible for mixture delivery, so the port's short-turn radius, the bowl, and the valve seat, combined with the chamber shape and the cylinder wall, are responsible for both the quality and direction of the flow.

SCHEMATIC ILLUSTRATION OF INTAKE/EXHAUST FLOW DURING OVERLAP

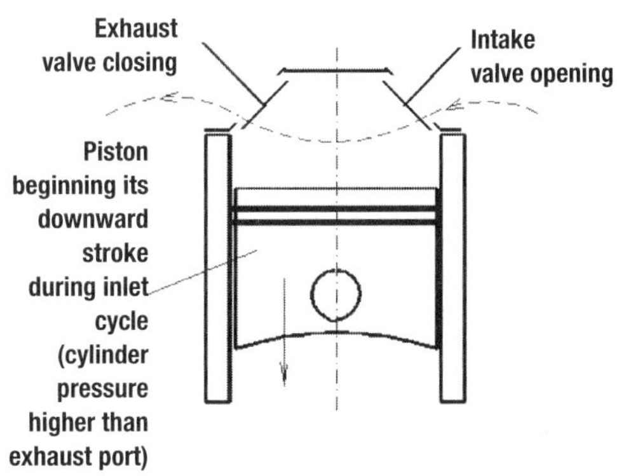

During valve overlap, a combination of fresh air/fuel charge and exhaust gas pass into the exhaust system. While the extent of this varies as a function of several variables (valve timing, overlap period, rod-to-stroke ratio, rpm, etc.), it is during this time that valve head and seat configuration is critical to overlap flow. It is of value to promote good low lift intake flow (during the overlap period) but disruptive to efficient cylinder filling if low-lift exhaust flow is excessive. Decreasing low-lift exhaust flow is important to preventing excessive fresh air/fuel charges from being lost into the exhaust path. Because cylinder pressure at exhaust valve opening is high, a reduction in low-lift exhaust flow will not be as critical as the improvement of low-lift intake flow, in the interest of increasing total volumetric efficiency. One flow bench method for evaluating this "cross flow" condition is to construct a dummy cylinder that will accommodate a piston, position the piston at its uppermost position (during valve overlap) and flow from intake through the exhaust port (with header pipe installed). By varying exhaust valve head and seat dimensions, low-lift exhaust flow can be adjusted to a minimum, thereby reducing the amount of over-scavenge during valve overlap.

CT: Do you think it's possible to generate too much mixture motion?

LW: Due to the fact that there are several components of differing masses in the flow (air at different levels of atomization, air and exhaust gas), excessive velocity can certainly cause separation and lost combustion efficiency. It's important to maintain reasonable mixture velocity in the targeted rpm ranges. This can be accomplished by developing artificially variable-geometry intake ports by "blowing" (directional changes) boundary layers.

CT: Which do you feel is more important, swirl or tumble?

LW: While each is a defined range of mixture turbulence, I've never seen evidence that tumble can deliver quite the same degree of total cylinder filling and mixture stratification that swirl permits. However, if the frequency of the tumble "loop" is properly configured and mechanically combined with the proper combustion space (head and piston), big power numbers with less specific fuel efficiency will result.

CT: How do you suggest the Saturday-night racer explore the fundamentals of improved mixture motion?

LW: Obviously, having the ability to actually study the head's flow characteristics (on a flow bench) would be advantageous. However, reading the combustion chamber, the piston, and the intake port color can reveal a lot. Dark intake ports are a sign that inert contaminants are getting back to the wrong side of the engine. Effective head-to-header steps combined with anti-reversion intake seat configurations (reverse-flow testing) should be adopted. Non-quenched areas that are clean in the combustion space should be modified by making their surfaces less vertical. You want to sneak up on this one with small steps at a time.

CT: Aside from engine or inertial chassis dyno analysis, how can the average racer evaluate the benefits of a fast-burn engine?

LW: Fast-burn engines require less spark timing and do less negative work than their more conventional counterparts. This is one reason they are capable of quicker response and transient acceleration (on the track). While measuring this on a dyno can often be difficult, using the actual race car for measuring elapsed times between two rpm (accelerating on the track) is relatively easy. Assuming you can reasonably duplicate traction conditions, the use of a g-meter will also do the job. Don't ever forget that the race car is the only dyno that really counts.

CT: What do you see as the future of fast-burn racing engines, in terms of areas where further improvements can be made to existing techniques?

LW: Work that I did in the '80s seems to have become part of the blueprint for today's high-performance and racing cylinder heads. I guess you could say it was a time of combustion revolution in those markets. During the last 15 years, we've (Endyn) developed and studied combustion spaces that will support seemingly impossible burn efficiency. There's a long way to go yet. Even though it may be a long time before multi-valve technology is accepted (with its greater valve area, better plug location, staggered valve size, and spark timing potentials) and allowed in circle-track racing, today's best will look like most of the old stuff from the mid-'80s. If the "perfect" combustion space is ever developed, all combustion chambers will look the same. And that's a fact.

ROUGHING IT UP

It's true that smooth intake ports help increase air and fuel cubic feet per minute into the combustion space, but there's more to the story. Note that the illustration shows two types of flow surfaces and their effects on boundary-layer conditions. Subsequently, this affects net flow. A non-mathematical definition of the boundary layer might be that zone or region in which velocity (of the moving fluid) is reduced to zero. The thickness of this layer depends upon variables that include the viscosity of the fluid (the greater the viscosity, the thicker the boundary layer). If dynamic pressure conditions are created that cause the boundary layer to separate itself from the flow surface, a low-pressure region develops. In a wet-flow environment, this is where liquid fuel tends to collect or "puddle." In an engine's intake track and combustion space, the degree and location of the puddles depend on issues beyond the scope of this story. But among them is the influence of flow surface texture, and that is worth discussing here.

ILLUSTRATION OF PORT SURFACE FINISH & EFFECTS ON BOUNDARY LAYER

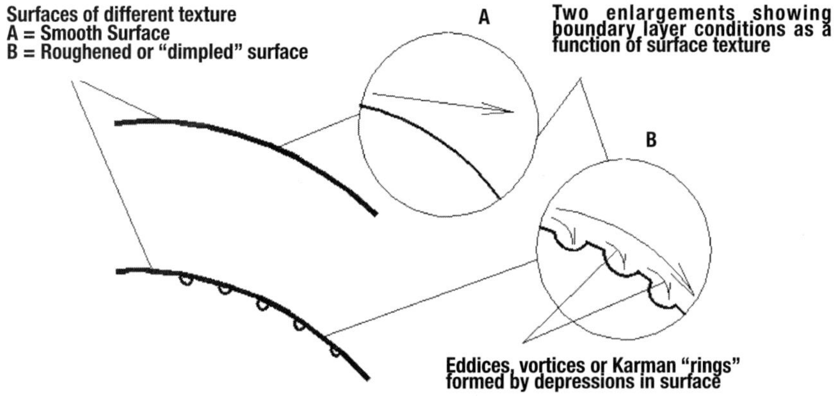

Surface "A" is smooth, allowing boundary layer separation and diminished flow efficiency of the passage. Surface "B" is intended to be rougher (or dimpled, as shown) to promote improved boundary layer control. The production of eddies increases low-pressure conditions at the flow surface, improving adherence of the boundary layer to the flow surface. A roughed surface creates a similar condition, enabling flow directional changes to be more efficient (less likelihood for turbulence at the flow surface).

By using a roughened (or dimpled) surface, the boundary layer separation is reduced, which enables a more efficient use of the flow path. Where a change in flow direction occurs (and this is important), a textured surface is valuable to both wet- and dry-flow systems. Though fuel may not be present, the air tends to behave like a fluid, responding to boundary-layer treatment and improved net flow as it would if fuel were in the stream.

Even in exhaust ports—especially in the short-turn area—roughening the flow surface can have beneficial effects, particularly during the closing side of the exhaust event. Failure to properly address the influence of flow surface texture can lead to ports that behave as if they are smaller than they actually are.

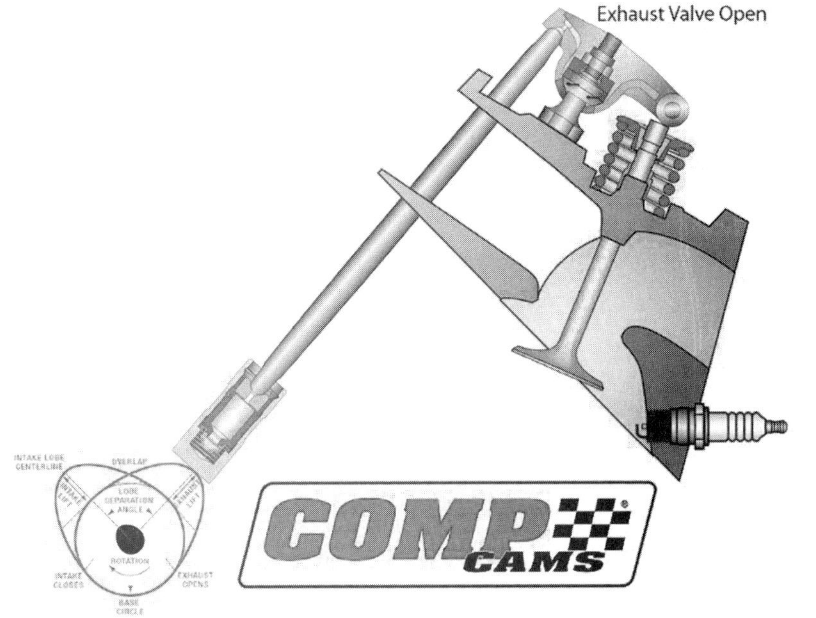

Chapter 12

VALVE TRAINING

Fundamental Cylinder Head Features, Functions and Modifications

Text and Illustrations by Jim McFarland

This chapter presents information that can lead to improved engine performance and reduce the cost of building and maintaining a competitive race package, thus separating winning from losing. So important is this message that we've elected to share an appropriate quote from Smokey Yunick.

"Unstable valve motion affects net cylinder pressure [power], reduces parts life, gets damned expensive, and loses races."

While the simplicity of this statement is often overlooked in the building of a racing engine, the complexity of a solution comes down to a companion rule of thumb: The only difference between a difficult problem and an easy one is knowing the answer. On this basis, let's dig into the basics of valvetrain dynamics, some of which step outside the normal boundaries of this type of story.

In order to compress years of experience and practical application into a short space, we conferred with two major players within the valvetrain parts community: Comp Cams' Thomas Griffin and Billy Godbold and T&D Machine's Larry Tores. The following are their responses to some application-specific questions, sprinkled with additional comments based on theoretical, personal, and unbiased hands-on perspectives. Bottom line? The core of this story is directed to the Saturday night racer.

Q: What are the most critical elements of a stable, reliable valvetrain, and why?

A: First of all, it's important to understand that valvesprings absorb and deliver energy, thereby liberating heat and experiencing a variety of harmonics that must be addressed in their design, construction, installation, and maintenance. Actually, they all wear out, but to extend their life and achieve maximum performance during this time, proper selection and installation is among an engine builder's primary objectives. Assuming spring manufacturers know their products, recommended selection and installation procedures are an excellent place to start.

It's also important to consult with a camshaft manufacturer to create a match between the camshaft lobe profile and the companion springs. According to Billy Godbold (Comp Cams), this combination is critical for several reasons.

"The camshaft and valvespring must work together," says Godbold. "A camshaft is the 'driver' of the system. A valvespring should store and release energy in a fashion that keeps the system (i.e., valvetrain) stable. Every cam profile inputs different harmonics into its valvespring, and every valvespring has certain harmonics based on design, material, and method of manufacture that it doesn't like. In good combinations (i.e., lobe profile and spring), the cam's energy inputs don't match where a valvespring is weak, anywhere in the rpm range. In other words, an unhappy situation can occur between lobe and spring if bad spring harmonics occur at an rpm where the lobe is delivering particularly demanding energy into the spring."

Q: How can you tell if an unhappy situation has been created?

A: "Telltale signs of an unhappy combination would be retainers and locks that show excessive wear, 'holes' in the torque curve (at some point in the rpm range) that do not respond to header or intake manifold changes, and broken valvetrain parts," says Godbold. "The valvespring and lobe profile can both be outstanding individual components, but if they don't like each other, an unhappy valvetrain package can result."

Tores' response to this question is the following: "First, and probably most important, is how the rocker-arm assembly is mounted to the cylinder head. The rocker-arm mounting bar must be secured to the head so no movement is permitted.

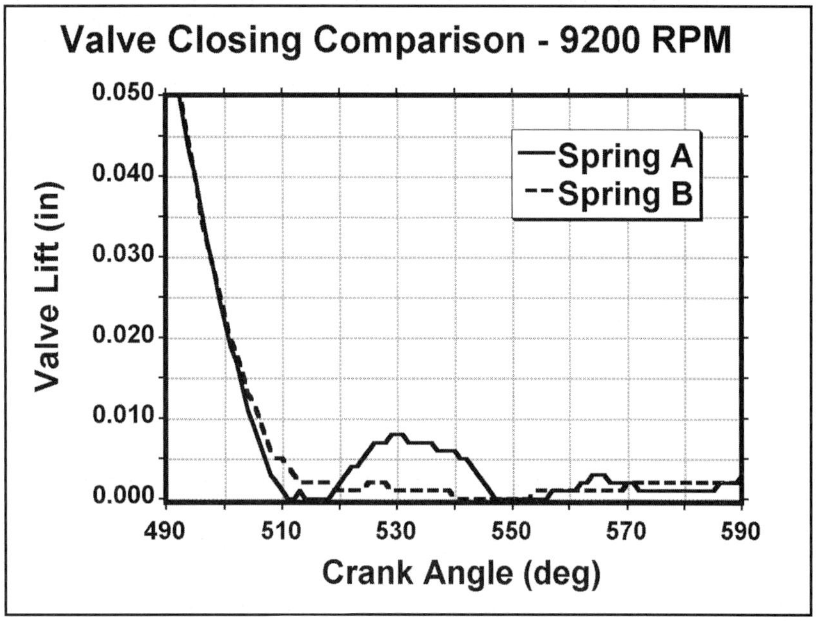

This data compares the valve control capabilities of two different valvesprings. Note the periods of "bounce," or lack of closing control (crank angles from about 520 degrees to 550 degrees and 560 degrees to 570 degrees). For intake valves, a condition of bounce (or "float") left unresolved allows cylinder pressure to be lost at the end of the inlet cycle.

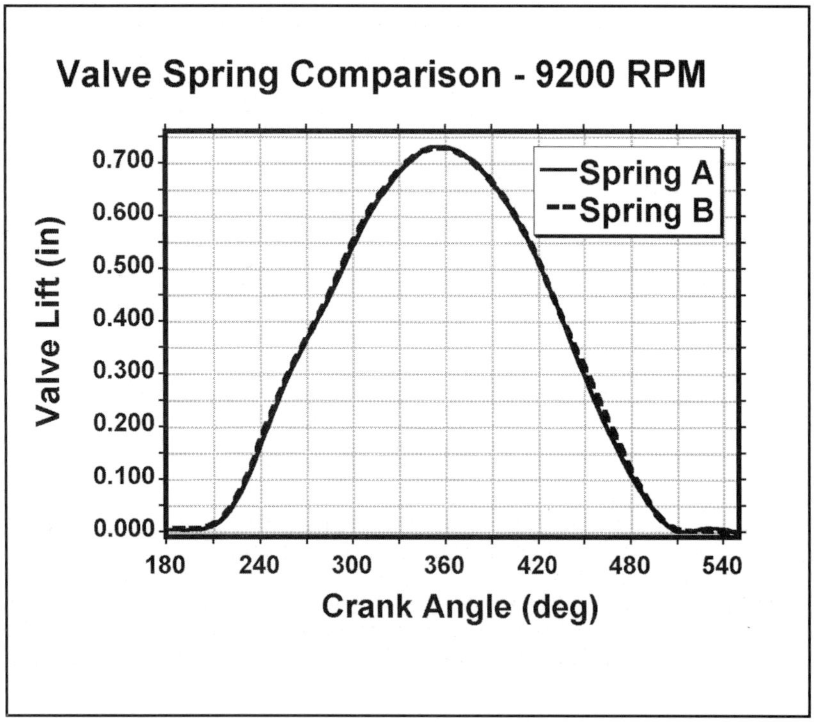

When comparing valvesprings on the Spintron fixture, the inability of a given spring to allow the follower to track cam lobe profile and control valve motion is often apparent at the point of valve seating. These two Nextel springs (evaluated at 9,200 rpm) show one spring more able to control seating action than the other (as noted between 510 degrees and 540 degrees of crankshaft angle).

In turn, the rocker arm and its shaft must be secured to the mounting bar rigidly, with no movement permitted. Rocker-arm mounting bars comprised of two or more elements are generally not as rigid as a one-piece bar. Most importantly, the one-piece, solid, stable base allows the rocker arm to do its job with no outside influence from an unstable mounting system. With a system like this, there is no wasted motion to compensate for.

"Further, lightweight, rigid rocker arms with very low-mass moment of inertia values allow valvesprings to control the valve and rocker arm with less pressure and/or to a higher rpm. Plus, rocker arms with a low-mass moment of inertia allow camshaft profiles to be somewhat more aggressive and net more area under the lift curve (for increased volumetric efficiency and torque) and still be controllable."

Q: How important is it to establish and maintain proper rocker-arm ratios?

A: This can be a tool. It is a method to not only change valve motion and create more or less aggressive movement, but also a means to compensate for (or help balance) the relationship of engine airflow, intake to exhaust. In the latter case, inadequate port flow or rules that don't allow port modifications can sometimes be addressed by increasing rocker ratio in addition to the use of camshafts with valve duration that favors the weakest flow values, typically the exhaust side of a given cylinder head.

According to Godbold, "Changing rocker-arm ratios is one of the easiest ways for engine builders to modify cam-profile aggressiveness. By increasing or decreasing the ratio, it's possible to make profiles quicker with more lift or slower with less lift. Typically, cam profiles are designed to run within a range of rocker ratios. For example, if a 1.5:1 profile runs well with a 1.8:1 ratio, chances are this profile was a pretty weak 1.5:1 design. More often than not, a good design will go 'ballistic' if the ratio is increased by more than one point above optimum.

"Hence, there is probably very little reason to employ ratios that vary less than half a point unless there's a desire to push the system to its dynamic limit."

However, as is the case with some aspects of engine building, it's possible to put more effort into a process or method than is necessary. States Griffin, "It's not only difficult to nail down ratio measurement to the third or fourth decimal place, but because of manufacturing tolerances, the Saturday-night racer can also wind up spending so much time building his engine that there's not enough time to enjoy his racing. Realistically, even the professional engine builders don't spend massive amounts of time checking rocker ratios for three-place accuracy."

From a slightly different but valuable perspective, Tores believes "rocker-arm ratios can be used as a tuning tool much like ignition timing, a camshaft, or intake manifold design. For example, different rocker ratios can be used on end and middle cylinders of a V-8 to provide power peaks at different rpm for different cylinders, resulting in a wider range of power.

"Ratios vary from about 1.5:1 to perhaps 2.3:1, depending on engine configurations and type of camshaft (flat tappet or roller tappet) and carburetor or restrictor plate size.

"One caution is worth mentioning. Very high-ratio rocker arms should be used on camshafts specifically designed for these ratios. Otherwise, very high valve opening and closing rates can be extremely damaging to valvetrain components."

Q: What are the best methods for verifying valvetrain geometry?

A: Once again, as pointed out by Griffin, it's possible to become over-involved in determining geometry. The so-called "mid-lift" approach is practiced by many builders in an effort to match operational and design ratios. Determining perpendicularity between the rocker tip and the valve stem involves noting when the rocker "sweep" stops moving in one direction and begins moving in another. Godbold says, "We typically recommend the mid-lift method to maintain the ratio as close as possible to theoretical [design]. The valve end of the rocker is perpendicular when it stops sweeping out and starts sweeping back toward the rocker."

In this same context, Griffin says, "On our Spintron, we maintain a certain 'sweep' across the valve for most of an optimization test session. After completing all optimizations, we raise and lower the rocker-arm stand to change the way the rocker moves the valve. Sometimes this will have a positive or negative effect on valvetrain control. This is used as a tuning method to deduct a few more thousandths of valve bounce or maybe gain a couple hundred more rpm. For engine builders who don't have access to a Spintron, we recommend monitoring how the valve sweeps, as shown in our catalog. This helps keep the builder from getting into too much trouble and keeps the rocker arm on the valve."

T&D Machine Products recommends that the rocker arm be perpendicular (i.e., 90 degrees) to the valve at mid-lift. "This configuration ensures the maximum amount of cam lift motion is transferred to the valve through the rocker arm," Tores said. "It also ensures the roller track or contact pattern on the valve tips is the minimum possible. For

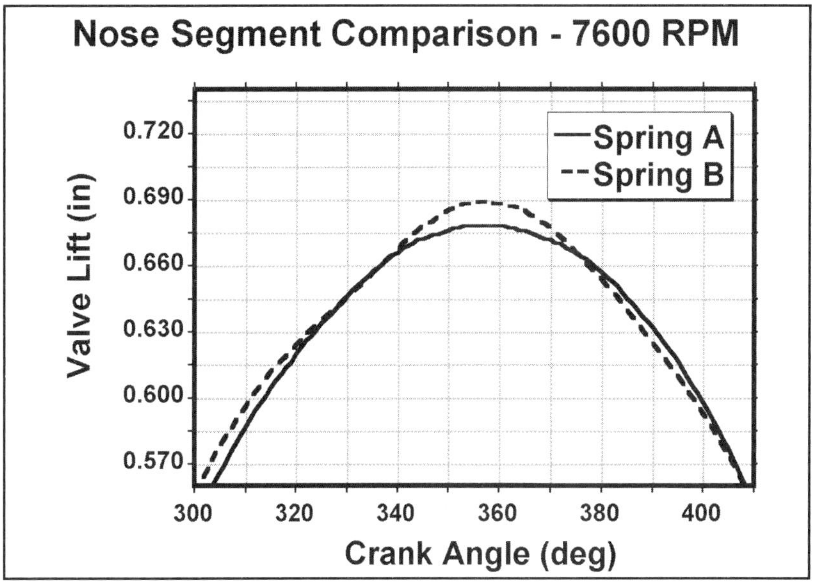

Valve follower–lobe separation over the lobe's nose can be affected by valvespring design, as shown in this comparison between two different Nextel Cup springs at 7,600 rpm. Sometimes called "pitch," this motion can provide brief gains in volumetric efficiency, although care should be taken not to allow the condition to lead to instability of the overall valvetrain.

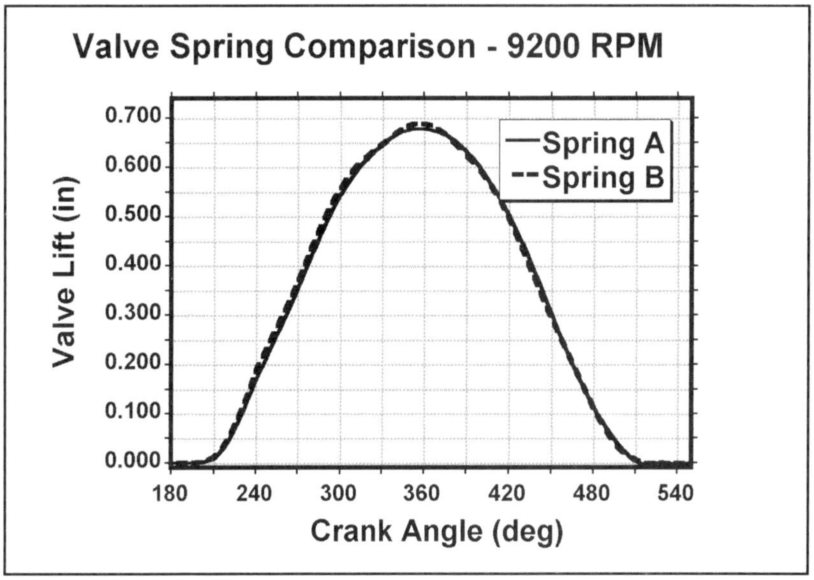

Once again, here is a comparison of two different valvesprings, one exhibiting more control of follower motion than the other (note nose separation between 330 and 360 crankshaft degrees). This data suggests the importance of selecting valvesprings that are compatible with the camshaft lobe profiles in use. Stated another way, the matching of spring characteristics to lobe profile is critical to valvetrain stability and optimization of pre-combustion cylinder pressure.

Comparison Between Pushrod and Simple Wire-Wound Spring

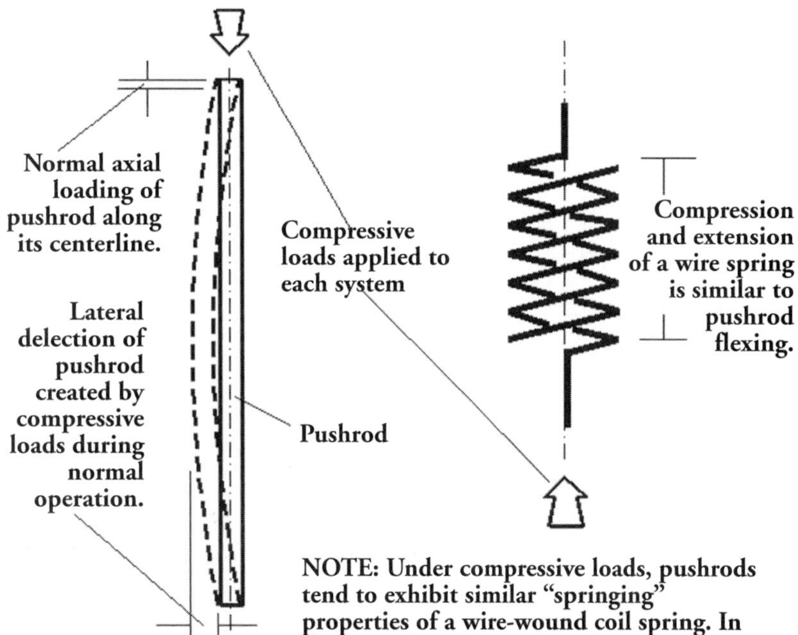

NOTE: Under compressive loads, pushrods tend to exhibit similar "springing" properties of a wire-wound coil spring. In the case of a pushrod, length changes created by normal loading tend to alter valve timing characteristics dictated by camshaft lobe design and rocker arm ratios.

example, a typical rocker arm for a small-block Chevy with 0.650-inch net lift will have a roller track of 0.037 inch. Mounting the rocker system very high or very low will produce an excessive amount of horizontal movement across the valve tip instead of the vertical movement that will move the valve. Not only is this wasted motion, it also results in pushing the valve tip toward and away from the rocker arm, contributing to unstable conditions in the valve/spring/retainer package."

Q: What are the most frequent causes of valvespring failure?

A: There is no rocket science here. Griffin is emphatic on this issue. "Misapplication and abuse!" he says. "Some springs fail because of material flaws or nicks in the material (i.e., stress risers) caused by mishandling. But the number-one cause for failed springs can be stated like this: 'The applied stresses exceeded the strength of the material and the residual stresses.' Basically it's like this: We sell springs and other parts, and we hope the engine builder and racer have enough sense to communicate with each other. This way, the builder can assemble an engine that will tolerate the abuse of its driver. Even in the best of circumstances, drivers don't always comply with the operating practices he's promised. Missing a shift or driving harder because of track conditions are subjects that don't always surface in conversations when parts fail. Drivers often avoid these topics."

"Other than over-revving an engine," Tores says, "most valvespring failures seem to result from flaws in the wire used to manufacture the springs. A valvetrain that is out of control can accentuate this. This can result from conditions like spring surge, where the valvespring reacts negatively to its natural frequency, or a condition where engine rpm is excessive for the valvespring pressure cam profile being used. Excessive forces like these acting on the valvespring can result in premature failure."

Q: Are there any on-track benefits from mixing rocker ratios? If so, on what basis can these choices be made?

A: While this subject can become a bit sticky, there are some issues worth considering. In particular, recent explorations in "single cylinder tuning," whereby each of an engine's cylinders is optimized for power, mixing rocker ratios to compensate for airflow inequalities or deficiencies is another engine builder tool.

Recent development of ignition systems that enable individual cylinder spark timing has added a reason for examining the power output of each cylinder. But even if this level of tuning isn't employed, there are other reasons for mixing ratios.

For example, Godbold suggests, "We like to see more rocker ratio on the intake lobe than the exhaust. This works especially well when the lobes are of similar lift and aggressiveness. Due to the smaller valve diameter, the exhaust port flow typically flattens out at a lower valve lift than the intake. Also, having the exhaust valve close slower allows it to spend a little more time near its seat during overlap. Plus, during this part of the closing cycle, the low valve lift helps increase exit flow velocity and provides additional signal to the incoming intake charge."

What's important to remember is that an engine is a package of components, each of which must work in harmony with the other. An example of this is the "cross talk" that goes on inside a single-plane intake manifold. Pressure excursions and even combustion residue passing into the manifold during early intake cycles can influence combustion efficiency of all cylinders, influenced by factors that include the firing order. Griffin and Godbold both have comments in this area.

"Since a V-8 engine is comprised of eight 'engines' that share a few common pieces (such as crankshaft, intake manifold, exhaust, and carburetor)," says Griffin, "there is plenty of room to tinker around with rocker ratios and cam designs. The engine is a system of mechanisms that must be optimized, then the effect of these mechanisms on the total system has to be considered so further efficiency and effectiveness can be achieved."

"If an engine responds to less rocker ratio on the exhaust, for example, you might want to try either shorter exhaust duration and/or a softer exhaust profile on another camshaft," adds Godbold, "but be careful when setting up pistons to provide clearance for the largest rocker ratio to be used. Engine builders will find piston-to-valve clearance is far more dependent on duration than rocker ratio and lift, but the change in rocker ratio still needs to be considered."

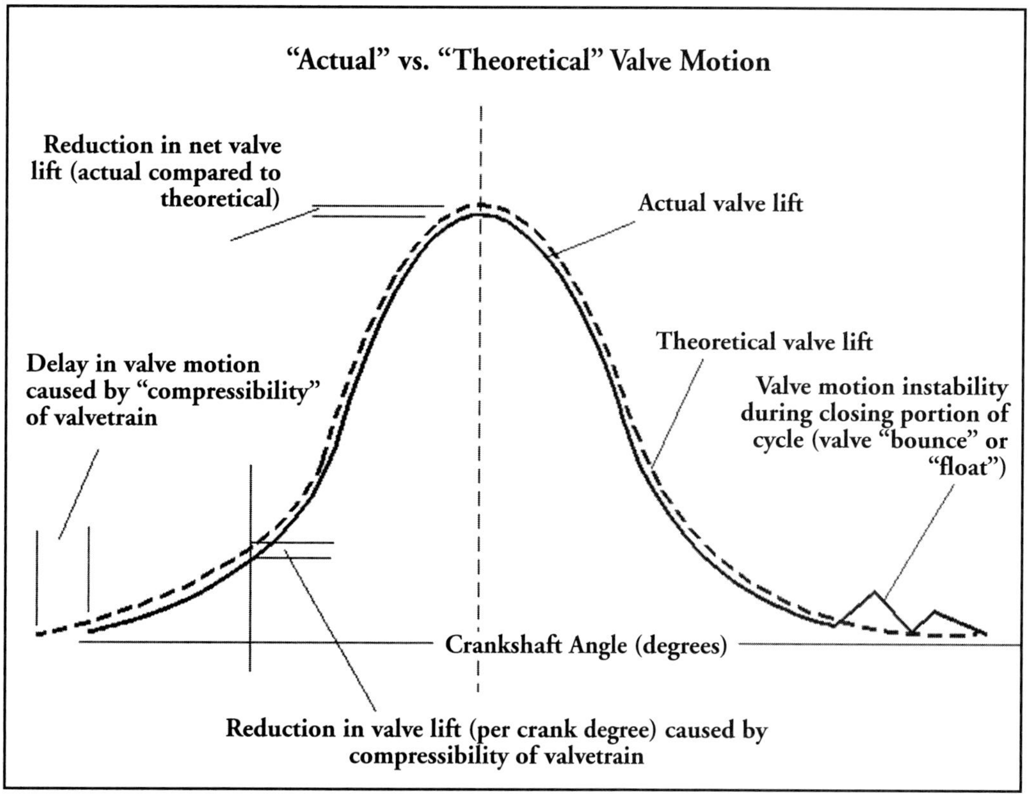

To further explore the possibility of using multiple rocker ratios on the same engine, Tores says, "Mixing rocker-arm ratios can produce different torque and horsepower peaks. Having these multiple power peaks will produce a wider average powerband. This usually results in more useable power on the track. A good starting point for selecting rocker-arm ratios is the camshaft manufacturer's recommendation.

"A higher numerical ratio should result in greater airflow and a higher rpm power peak. Typically, the combining of different power peaks from different cylinders would offer a greater amount of power at lower engine rpm (i.e., to help accelerate a car out of a corner), at the same time offering more top end to achieve greater straightaway speeds."

Beyond these comments, changing rocker ratios can also compensate for cylinders with inadequate port flow. In that regard, the next question addresses the issue of how port flow can be effectively increased by rocker-ratio changes.

Q: With some Nextel Cup engines now using multiple lobe profiles on the same shaft, how can rocker-arm ratios be selected accordingly?

A: We've covered ways to address how cylinders talk to each other in single-plane intake manifolds and collector-joined exhaust systems. Various and predictably varying pressure excursions can influence intra-cylinder combustion efficiency and power output, especially in the intake system.

Stated another way, some cylinders receive more air and/or different air/fuel charges than others. Among the causes are inequalities in cylinder-to-cylinder airflow created in a running engine, despite efforts to calibrate or equalize flow during air bench tests. As a result, some cylinders make more power than others. While the ability to identify and confront this problem typically requires precision measuring equipment and a fair amount of analysis experience, Engine Cycle Analysis or ECA is gaining in use and popularity among the more aggressive race teams and engine builders.

Even though some of the following responses to this question may appear a bit off target, there's clearly some between-the-lines information.

"Most, if not all, of the Nextel Cup teams have very extensive dyno testing programs as well as on-track testing," Tores says. "Part of the tuning process for a given racetrack, length, banking, and so on will be the rockers, as well as valvesprings and valves. Usually, the results of Spintron and dyno testing will dictate which rocker-arm ratios are used: more aggressive for a qualifying engine or configuration and more conservative for the longer race." Obviously, engines for qualifying are not the same for racing.

Griffin responds differently. "Cylinder pressure analysis (ECA) helps to understand the interaction

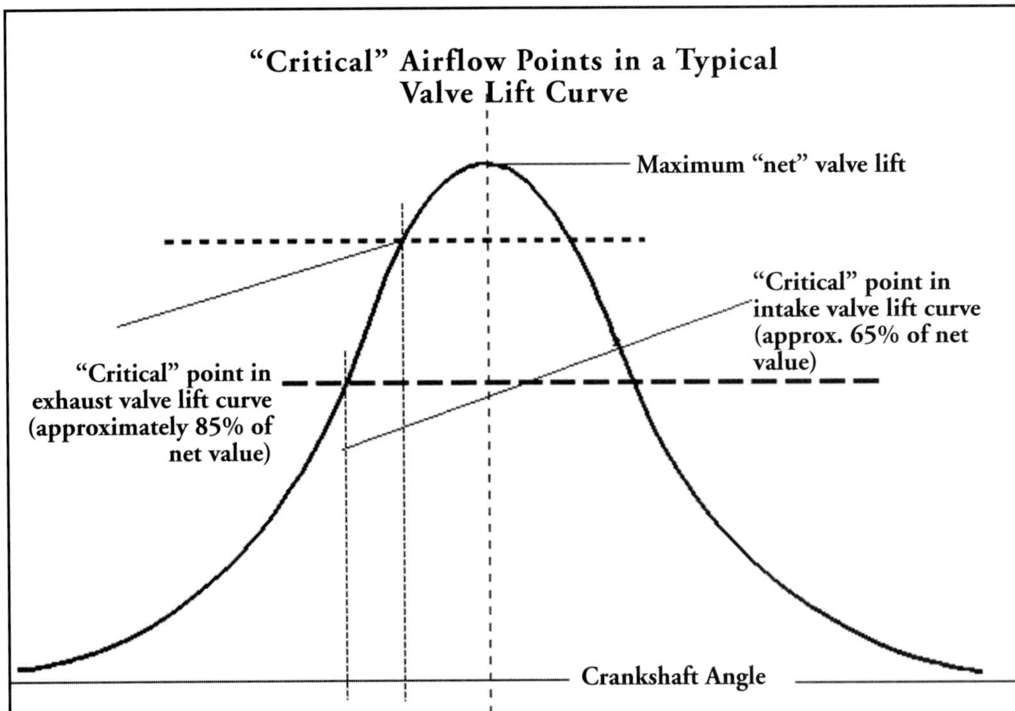

NOTE: These two points are identified as "critical" in the airflow analysis of intake and exhaust port flow, as it relates to net valve lift values for intake and exhaust cycles. Notably, each point represents that stage of the lift cycle (intake and exhaust) at which each valve resides for the greatest period of time...on the opening and closing sides of the profile. These so-called RMS (root mean square) values are particularly helpful in determining airflow efficiency of ports.

of the mixing of cam profiles on overall performance," Griffin says. "Seat-of-the-pants determines how well any optimization routine works." This agrees with the prior suggestion that individual cylinder performance follows individual cylinder analysis. Nothing new here. Reading spark plugs is an age-old technology. However, in-cylinder pressure analysis as a function of crankshaft angle (ECA) brings the measurement technique inside an operating cylinder to help identify weak or strong cylinders. Then look outside the cylinder to take steps that could include multiple profiles and rocker-arm ratios.

What Godbold said falls right in line with this reasoning. "By looking at in-cylinder data (e.g., by experimenting with individual cylinder rocker ratios), engine builders sometimes find certain ports are weaker than others," Godbold says. "The weaker the port, the more duration and area under the lift curve the cam needs to provide to make power at a given rpm. Every profile Comp designs is intended to work within a range of rocker ratios; you cannot just keep adding more rocker ratio to weak ports.

"Typically, you should run the maximum safe rocker ratio on a given profile. However, some engine builders have found increasing the duration on the more restrictive ports can help smooth out the power curve or increase peak power. Most importantly, the larger the difference in port flow from cylinder to cylinder, the greater the potential to find improvements with these techniques." Remember the last sentence.

Q: What are the structural reasons some rocker arms break and some don't?

A: According to Griffin, "Rocker arms are loaded [during normal operation] in bending, which is the most horrible way to load a structure. By making these components out of improved materials with stiff cross sections, high-rpm loading can be less detrimental. Failures also occur from high-stress points in the threads where the adjuster nut is positioned. Since the rocker arm is bending, the surface is being put into tension. As a result, shot-peening has helped increase the fatigue life of aluminum rockers because of the compressive stresses imparted on the surface of the body of the rockers by this process."

As you might expect, material selection is critical to parts life. Aluminum is not just aluminum, no more than parts are parts. Tores is particularly clear on this point. "There are a variety of reasons that one rocker-arm design works better than another," he says. "Selecting the proper alloy incorporating high strength at elevated temperatures, low notch sensitivity, and high levels of fatigue resistance are important factors.

"At T&D, we use SAE 2024 aluminum, the most fatigue-resistant of all the hard-alloy aluminums. It also has low notch sensitivity, which is an indicator of how readily a particular material will start to fracture under high stress. In the final design of a rocker, sufficient material should be used in critical areas to ensure a high safety margin.

"Generally, rocker arms do not fail from tensile strength failures. They usually fail from fractures in highly stressed areas such as the radius under the nose or in the rear of the rocker arm, or a cut for a valvespring. Failures can result from these radii being too small and not properly distributing the stress evenly to the areas adjacent to the radius. Rocker-arm failures can often come from alloys that do not have good elevated temperature strength. Remember, a race engine may operate at temperatures up to 250 degrees F, and these temperatures can cut the strength of some aluminum in half."

From a slightly different perspective, Godbold says, "Rocker-arm failure is typically an alarm for an engine builder to look for problems in other areas. Aluminum work hardens and becomes brittle. Hence, aluminum rockers should be treated with wear limits and replaced at regular intervals. If you're using aluminum rockers that were hand-me-downs or ones with excessive use, a failure should not be a surprise. However, if a rocker fails early in its life, pay careful attention to wear of the retainers, locks, and/or valve seats. It is quite likely that valvetrain instability may have contributed to the failure.

"Comp Cams prefers steel rockers to aluminum rockers in any application where steel is available. Steel rockers should only fail if they are used outside their design application, if sufficient care was not taken with design, or if they are improperly modified. In fact, I cannot recall seeing a Comp Cams full-roller steel rocker fail that had not been modified by the customer. Even in steel rockers, bearings should be replaced periodically.

"Some classes of racing require the use of stamped steel rocker arms. The level of competition in many of these classes often results in the cam forcing the pushrod through the pushrod seat of the rocker. This type of failure is very difficult to resolve without choosing a cam profile that makes less power."

Q: How can you equate Spintron testing to actual engine operation, given the fact valvetrain components aren't operating in the changing environment of cylinder pressure?

A: Historically, there are problems associated with relating simulated conditions to operating conditions, almost independent of the science involved.

It's the nature of modeling complex events by whatever method employed, be it via computerized simulation or the use of laboratory devices. In the case of spinning valvetrain components and comparing stability and durability performance with that of a running engine, there are issues to be considered.

Countless hours of Spintron work have caused Griffin to draw certain conclusions. "Oddly enough, valvetrain failure rates occur close to the same number of cycles, Spintron to dyno," he says. "Besides the lack of cylinder pressure, engines do not experience the input of combustion heat and its affect on the visco-elastic properties of the valves and seats. As a material's temperature changes, its stiffness changes. Along with that, different materials will have different changes in these properties. It is entirely possible that we never see this effect, and it is also possible that the effect is minimal. We just don't know!"

In this same context, Godbold has some additional points. "We compare dyno results and other engine data to validate much of our Spintron data," he says. "While some differences appear when making direct comparisons, we've seen almost identical pushrod strain data on the Spintron and a running engine. There's always the possibility that cylinder pressure at exhaust valve opening is substantial. However, the pushrod loads at exhaust opening on the Spintron sometimes exceed 2,000 pounds. Hence, forces to overcome the inertia of valvetrain components are most likely much higher than any force due to cylinder pressure.

"On the intake side, pressure differences should be zero at intake closing and small at intake opening. In most applications, we see the same characteristics on the Spintron and a running engine. Other factors, such as valvetrain harmonics and pressure differences across the valve in a running engine, tend to create data shifts from one measurement environment to the other. Overall, there are significant benefits when comparing Spintron information to that obtained from a live engine."

Cycle testing of rocker arms is another viable method of evaluation as a companion to Spintron analysis. In this regard, according to Tores, "We test rocker arms on a hydraulic load machine. At loads of about double the typical valvespring pressure, we increase the load and cycle frequency until failure is induced. The 'cycle to failure' numbers mimic that of race conditions. The most critical part of testing is the operating temperature, and this can be duplicated on the Spintron. We feel only a small percentage of the load on a rocker arm is from cylinder pressure."

Q: I'm a Saturday-night circle-track racer building my first serious engine. What guidelines should I follow when selecting and installing valvetrain components?

A: Caution and good judgment is a pretty common thread on this topic. Even the three experts made comments that fall into this line of thinking. "Every application is so different that very

Short-Side Reversion Paths in Intake and Exhaust Ports

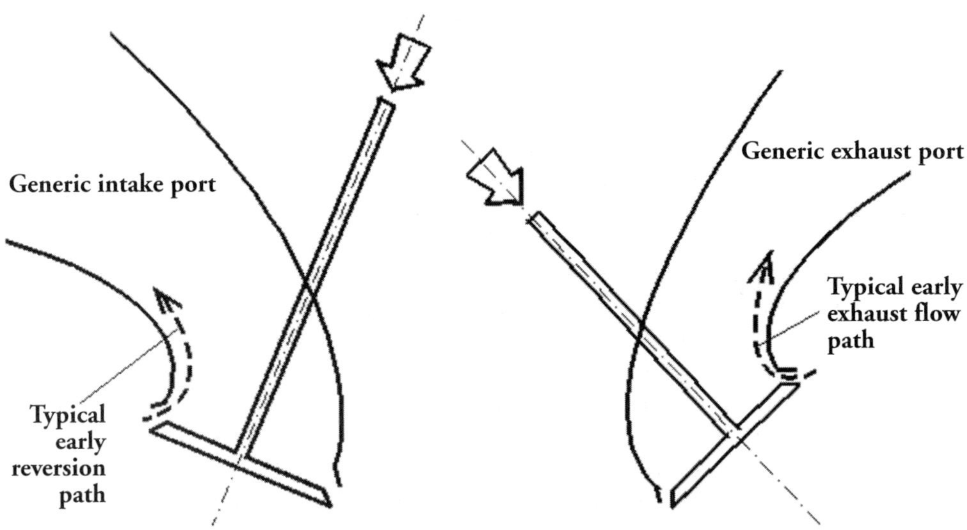

NOTE: Of the methods used to "fine tune" a valvetrain (camshaft profiles matched to low-lift port flow characteristics), comparing "reverse flow" potential on an airflow bench is often worthwhile. By comparing short-side airflow (both intake and exhaust) toward and away from the valve seat, valve seat pocket angles can be determined that affect "forward" flow and "backward" flow. Camshafts that provide slow off-seat intake lift rates and rapid off-seat exhaust lift rates become keys to helping damp reversion into the intake port and aiding early exhaust flow, especially during the initial stages of the blow-down cycle.

little can be held as true across the board," Godbold says. "It's wise to establish a relationship with someone you can trust at a major camshaft company. While you may experiment with other vendors later on, that primary relationship can be extremely helpful toward getting a competitive engine built that will stay together. The second piece of advice is to not let your first few engines become an R&D project for the cam company. Make sure some variation of what you are running has been used successfully in someone else's engine.

"Often someone will use a computer desktop engine program to select the perfect cam specs that have some combination of a big intake, a small exhaust, or a really wide lobe separation. Going off the beaten path may be necessary in the long run, but start with something known that can win races. Then go for the experimental parts later. Going the other way around can lead to serious confusion as you try to figure out which unique component of your combination is giving you fits on the track."

Tores doesn't like tricks, and he made that point pretty clear. "Keep it simple," he says. "Save the tricks for later. Allow some room so that other things can be tried, such as different rocker-arm ratios. Don't go out and buy the latest trick-of-the-week combination. Durable parts are more important. If nothing fails after a period of racing, then new components can be tried, following a scientific method of analysis."

"Listen to the old guys," Griffin says, even more to the point. "Go conservative, but ask questions about what people have done to really screw things up. If some people don't want to respond to your questions, go ask someone else. The non-responders might not have any real answers anyway." Simple enough.

Q: Are lightweight valvetrain components valuable?

A: If there is value, which parts provide the greatest benefit? There were multiple answers, and the perspectives vary slightly. "The closer to the valve, the better your money is spent on lightweight components," Godbold says. "Stiffer is almost always more important than lighter as you get to the pushrod and camshaft core. Rockers typically like stiff just as much as— if not more than—light. Get the valve as light as possible and the pushrod as stiff as possible; work out a good compromise everywhere else.

"Heavy stainless steel valves with steel retainers and standard 5/16-inch pushrods make one of the hardest valvetrain combinations to control at high rpm. The heavy valve system increases pushrod flex (it acts like a spring). When the pushrod springs back, it can surge the valve and valvespring. Then, when it comes back together on the closing side, the pushrod can flex again, resulting in valve closure well before the tappet gets to the soft part of the closing ramp where rates are controlled. Heavy systems are greatly limited to how hard you can push the components, both in terms of rpm and aggressiveness of useable profiles."

"Components that will benefit from weight reduction are the valve, retainer, locks, and even the spring," says Griffin. "The rocker arm benefits from having a lower moment of inertia—where lighter on a weight scale doesn't imply a lower moment of inertia—and a stiffer body. Pushrods benefit from having stiffer bending properties, a result of larger diameters, tapers, and thicker walls. Another aspect of pushrods that has been examined is the bending natural frequency, which can be increased by making the diameter as large as possible and the wall as thin as possible. Little testing has been done to support any theories."

"Lightweight valvetrain components are very important," says Tores. "But always remember that you have to be able to finish the race in order to win. So durability is equally important.

"In most Spintron testing, everything on the valve side of the rocker-arm fulcrum, or pivot, is critical. Any weight saved here will usually result in an increase of limit speed with the same valvesprings. Weight on the pushrod side of the fulcrum tends to be somewhat less critical because the velocities are so much less (i.e., less weight must be started, stopped, and restarted into motion). Lightweight retainers and valve locks, valves, and rockers are always the best targets for weight reduction as long as they are reliable. A part may be considered reliable if it lasts for one complete race, no matter how long or short. How practical this approach may be will depend on your racing budget."

FINAL NOTE

As in the case of most component selections, compromise is almost unavoidable. When one area of concern becomes optimized, another likely is not. Striking that elusive balance between maximum power and optimum durability often boils down to having adequate information upon which to base decisions. Past experience, especially if someone else paid for it and is willing to share, frequently helps avoid repeated mistakes. Seek reliable information sources, ask a rash of questions, and draw conclusions based on what you believe best fits your needs. It's usually a safe bet to assume the manufacturer of a part is a more knowledgeable resource than other sources.

Chapter 13

MANIFOLD BASICS

Taking a Closer Look at This Highly Influential Engine Component

Text and Illustrations by Jim McFarland

Dimples can be used to create small eddies in areas where liquid fuel forms due to mechanical separation with the airstream. These little kernels of energy help re-atomize liquid fuel, restoring air/fuel mixtures to a more combustible form and aiding net power. This Hi-Tech Tork-Link manifold uses dimples on the floor of both planes and all runners.

Stop right here. Don't jump into the meat of this story yet, because there are some fundamentals we need to discuss first. Although it's likely you've read about some of these elements, don't assume you completely understand them until you've finished reading the next few paragraphs.

Intake manifolds don't just connect inlet air to an engine's cylinder heads. If you're a student of the importance airflow quality plays in making raw horsepower, then recognizing the value of a properly designed and applied intake manifold is a given. But you may not have been educated in the subject yet. It is the ability of an intake manifold to provide sufficient airflow volume and quality that contributes to good combustion efficiency and power, whether the system is connected to a carburetor or a version of fuel injection.

Now consider the influence an intake manifold can have on equalizing air (or mixture) distribution to each of an engine's cylinders. Considerable effort has been and is being devoted to optimizing engine performance, one cylinder at a time.

Understand that intake manifolds which address the equalization of cylinder-to-cylinder airflow on a running engine (not on a flow bench) have become requisite to optimum performing powerplants, especially those used for multi-lap, circle-track racing.

Overall, the dynamics of an intake manifold have become a study unto itself, particularly when atmospheric pressure is the only source of aspiration. The ability to generate and control specific pressure excursions in an intake manifold system of runners is significantly governed by the physical dimensions of the design. This article is intended to reach into certain areas of intake manifold understanding that may be unfamiliar to you, but certainly of potential value in making horsepower, specifically where parts longevity is vital. Now you can proceed.

MANIFOLD TYPES AND THEIR CHARACTERISTICS

Regardless of an intake manifold's design, a common condition exists at the beginning of each inlet cycle. To visualize this condition, consider the following simplified events to be unfolding in very slow motion. At this point, our sample engine is a single-cylinder, four-stroke cycle version.

Let's start by assuming the intake stroke begins without any prior cylinder firing. Thus, when the inlet valve opens (piston descending), atmospheric pressure is higher than cylinder pressure. However, a volume of air already is residing inside the intake manifold, offering inertia (or resistance) to movement into the cylinder. The amount of inertia offered depends largely on the volume and density of air in the manifold.

Additional post-carburetor fuel conditions (atomization) can be accomplished by the texturizing of surfaces on spacers or adapters. These circumferential marks, scribed around the corners of this adapter, are intended to increase the energy content of air passing by the irregularities. This method is more effective in wet-flow systems than those only passing air.

Most intake manifolds (and intake ports) tend to direct air or air/fuel mixture to specific parts of a combustion chamber. Especially at high flow rates (high rpm), these patterns indicate areas of either air/fuel separation (wet flow) or high impingement of air (dry flow). In either case, strategically placed dimples can offset the effects of poor flow direction (caused by manifold and/or intake ports) by creating so-called Karman rings or eddies that activate boundary layer flow (that which is next to the surfaces), aiding fuel mixing and combustion.

As either volume or density (or both) increases, it's tougher to start and stop the airflow, which is exactly what takes place during inlet cycles: Air is started when a valve opens and stops when it is closed. Of course, the dynamics of this involve compressibility and elasticity of flow, but that's another issue. For now, we're just concerned about how quickly inlet air can be set into motion to fill a cylinder.

So, here's the point. The more air (or mixture) inside an intake manifold, the more inertia is offered to starting or stopping flow throughout a complete inlet cycle. If we want to sharpen throttle response by reducing intake flow inertia, a reduction in the volume of air required to begin filling a cylinder will help. The converse is true.

If an intake manifold is designed to have two cavities (volumes), it's commonly called a two-plane (at the risk of oversimplification). Turn this divided volume into a single volume, and the manifold becomes a single-plane. Therefore, it should not be surprising that two-plane manifolds tend to be more throttle responsive than single-plane designs. This, of course, assumes that the carburetor calibration is correct for the respective fuel metering signals of the two types of manifolds (single-plane manifolds tend to have weaker signals, all else being equal). Add a plenum divider to a single-plane manifold, and it will demonstrate throttle response similar to a two-plane.

Then, there's another issue when comparing these two types of manifolds. In a single-plane, during the inlet cycle of any cylinder, air is supplied through the carburetor and from the remaining volume of the manifold. That's correct. Air and fuel already resident in the plenum and runners of the manifold are subject to being inducted during a cylinder's inlet cycle. Portions of mixture from all other runners (and the plenum) can be added to air/fuel mixtures passing out of the carburetor. So, by virtue of their respective differences in internal volumes (open or divided), single- and two-plane manifolds experience mixture distribution variances, cylinder to cylinder and cycle to cycle.

REVERSION AND THE INTAKE MANIFOLD

First, let's establish what we're calling reversion. Sometimes this term is misapplied to what it represents.

At the instant of intake valve opening, cylinder pressure is typically higher than that inside the intake manifold, even though exhaust is still passing out of the exhaust port. Because of this higher-than-manifold cylinder pressure, a measure of combustion residue passes back into the manifold. Keep in mind this is not a combustible residue, so it may become a contaminant for fresh air/fuel mixtures entering the cylinder later.

For the next few degrees of crankshaft rotation, this backflow or reversion continues until the instant cylinder pressure, manifold pressure, and exhaust port pressure are equal. However brief this period, atmospheric pressure takes over and begins

Also associated with intake manifold (and port) flow discharge patterns is the combustion chamber floor, or piston tops. As pistons approach TDC and become aware of the combustion roof's (chamber) shape, mixture (or air) motion during this phase of combustion can be aided by dimple patterns as shown. The result enabled slight reductions in spark timing and an increase in indicated mean effective pressure (i.m.e.p.)—reduced work against an ascending piston during the compression stroke. We can determine where such dimples should be placed by reading combustion patterns on the tops of used pistons. Look for clean areas as candidates for dimples.

filling the cylinder. Generally, whatever flowed back into the manifold is now mixed with fresh air and fuel, comprising the next combustible intake charge.

Therefore, it is important to combustion efficiency (and power) that reversion be minimized, both in amplitude and duration.

Large runner manifolds, single-plane manifolds, or both tend to offer less reversion pressure resistance than smaller runner or two-plane manifolds. In addition, the lower inlet flow inertia offered by small runner or two-plane manifolds helps dampen the effects of reversion; so do late-opening intake valves and low back-pressure exhaust systems, but that's another topic.

If you discover exhaust-stained insides of manifold runners, it's reasonable to assume that the engine is suffering from reversion-contaminated combustion. And if the bottom of a carburetor shows the same signs, the extra fuel and timing you've been dumping into the engine was needed to cover up a massive reversion problem.

Finally, the phenomenon of reversion is never absent from an engine. It only varies as a function of rpm, all else being equal. As rpm increases, there is less time for reverse pressure to influence the intake system, but some measure of exhaust gas (comprising the physical elements of reversion) tend to remain in the combustion space at all rpm and loads. The important thing is to recognize this fact, and take whatever measures are available to minimize the effects of reversion, certainly within the span of rpm in which you're trying to optimize power.

FLOW RATES VS. PISTON DISPLACEMENT VS. RPM

These elements are unavoidably tied together, regardless of the type of intake manifold used (single-plane or dual-plane). While you might think these topics are concerns of a manifold designer, they come into play when matching manifolds to applications, including fuel injection manifolds, for which there are no plenums.

It is known to the art that a mean flow velocity (m.f.v.) of 240–260 feet/second is developed in an inlet path, closely matched to peak volumetric efficiency (torque) rpm. In fact, it is desirable to select dimensions that boost torque at preselected rpm. Since flow velocity is a function of the runner cross-section area (independent of length), piston displacement is related to the rpm at which this m.f.v. is achieved. For example, if a manifold runner with a 3.0 square-inch cross section reaches this desired m.f.v. at 5,000 rpm on a 350-cid engine, it will develop this same m.f.v. at a lower rpm on a 400-cid engine. As you would expect, the opposite is also true.

As a result, intake manifold selection should certainly include the rpm range in which it is designed to operate. More importantly, when choosing an intake manifold, the point at which the induction system will contribute to peak torque should be considered, given the fact that both intake and exhaust systems can develop independent torque peaks. A prudent step is taking measurements of intake runner section areas, which leads us to the issue of runner taper.

Aside from digging into the essentials of taper versus no taper, most intake manifolds incorporate tapered runners. In simple terms, you can measure the section area of a runner's entry, add this value to the section area of its exit, and average the numbers. This information will help you determine the approximate rpm at which the runner will reach its torque peak. In case you've missed previous inclusions of one equation that produces this rpm, it is as follows: peak torque rpm = [cross section area] x [88,200] / individual cylinder volume.

If we plug in some values for these variables [cross section area = 3.0 square inches, individual cylinder volume = 45.0 cubic inches], a calculated peak rpm point for a runner with this section area [even if averaged] is 5,880. You can reverse the process to determine what cross section area is required to produce a torque peak at a desired rpm by solving this equation for cross section area:

Cross section area = [peak torque rpm] x [individual cylinder volume] ÷ 88,200.

Using these same values [peak torque rpm = 5,880, individual cylinder volume = 45.0 ci], the cross section area equals 3.0 square inches.

Given this information to intake manifold runner sizing, it follows that an engine's torque curve can be manipulated by either the design or selection of runners that produce m.f.v. characteristics where we want them. Again, it's a matter of deciding before you make a manifold selection where an engine will be run relative to rpm. Don't choose a design simply because you believe it'll make more power.

Parts selection based on the intended rpm range is critical to getting power where it's needed. In regard to this subject, there's one other point to consider. Much like in the performance of exhaust header primary pipes, intake manifold runner length affects the amount of torque produced above or below peak torque. If length changes tend to rock a torque curve above its peak rpm, you should consider this aspect of manifold performance when selecting a design. However, since many manifolds don't have equal length runners, some compromise is unavoidable. But in the interest of tuning one cylinder at a time, it makes sense to at least consider runner length when matching a camshaft to a manifold. Remember, not all camshaft lobe profiles need to be of the same intake and exhaust timing patterns. (Matching lobes to runners is an emerging concept among some engine builders.)

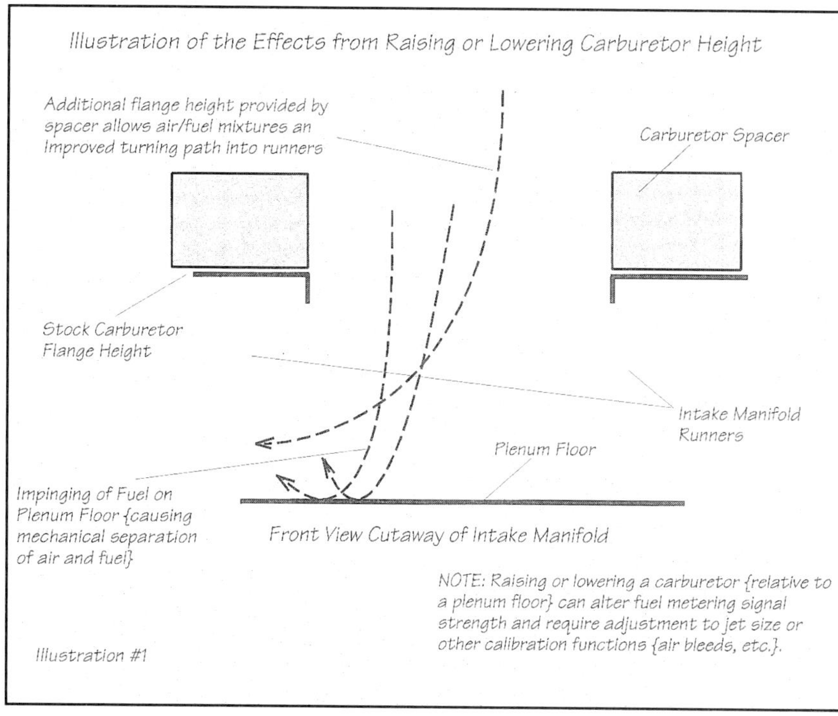

Illustration of the Effects from Raising or Lowering Carburetor Height

Illustration #1

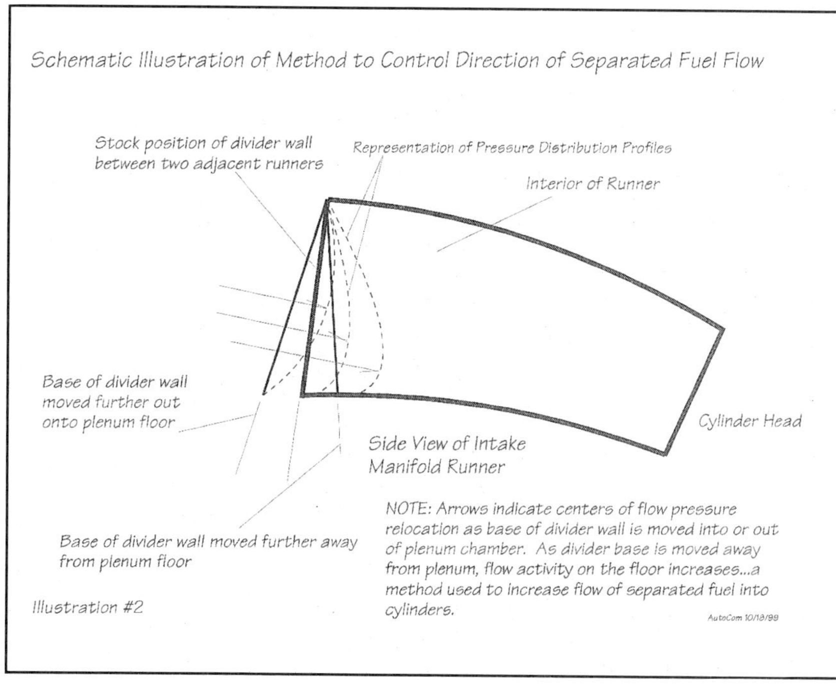

Schematic Illustration of Method to Control Direction of Separated Fuel Flow

Illustration #2

MYTHS AND PLENUM CHAMBERS

Nothing magical here. While it's possible to configure an intake manifold's plenum volume to provide a torque boost somewhere in an engine's rpm range, space limitations often prevent this option—but it can be done. A more frequent issue is how to design, use, or modify a given plenum's dimensions specific to a particular application.

In the vast majority of circle-track applications involving intake manifolds with plenum chambers, this cavity is used to redirect either air or air/fuel mixtures from an entry point into the runners. For example, in a single 4V carbureted V-8 application, air and fuel are required to move from basically a vertical direction into something that approaches horizontal. Because of its compressibility, air can negotiate this change in direction and energy more easily than fuel. Herein lies one of the fundamental problems in plenum design or modification.

If we conclude that peak torque flow velocity in an intake runner is in the 240–260 feet/second range, then flow rates approaching 600 feet/second in one throat of the 4V carburetor are not

Graph Showing Effect of Runner Cross Section Area on Torque versus RPM

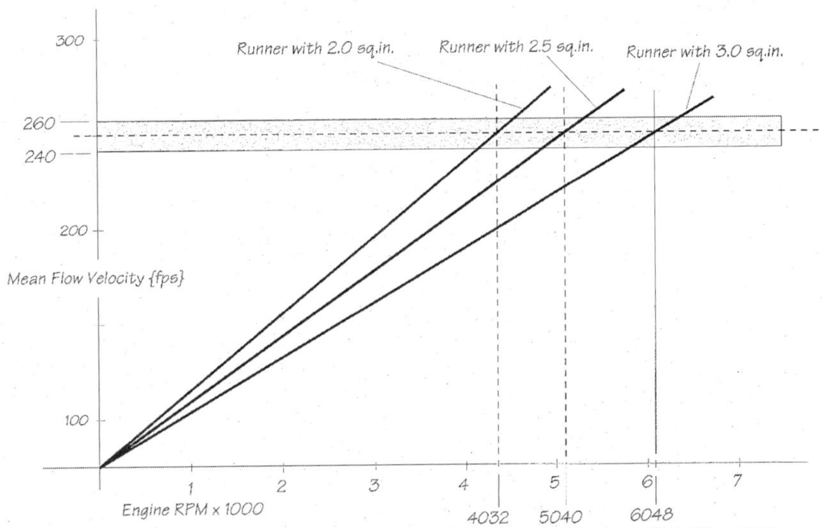

NOTE: This sample intake manifold has runners with three different cross section areas {2.0, 2.5 and 3.0 sq.in.}. Each reaches its critical value for mean flow velocity at three different rpm...representing the rpm at which each runner produces a torque boost.

Illustration #3

unreasonable to assume. At higher rpm, the rate is obviously faster. But even at peak torque, air and fuel can exit one throat of a carburetor at several hundred miles per hour, depending on the carburetor throat size. Now, considering the difference in the unit weight between air and fuel, it doesn't require much homework to visualize the difficulty in having both turn into a manifold runner while remaining mixed.

It is little wonder that the mechanical impingement of air/fuel mixtures against the floor of a plenum chamber [particularly the upper plane in two-plane manifolds] accounts for problems in maintaining mixture integrity in the runners. The problem becomes worse with increasing rpm and mixture flow rate.

One partial solution to this condition is the raising of carburetors via various types of spacers, preferably of the open variety. The object here is not to increase plenum volume for any reason other than to enable improved turning of wet flow from the carburetor base to the manifold runners, minimizing the mechanical separation of air and fuel during a sudden change in direction. Interestingly, the addition of a four-hole spacer does little to address the problem of air/fuel separation, although it can strengthen fuel-metering signals and sharpen low- and mid-rpm throttle response.

OPTIMIZING SHORT AND LONG RUNNER POTENTIAL

The issue of tuning one cylinder at a time can be applied to intake manifolds. In this case (all else being equal), long runners tend to improve power below torque peak, while shorter runners boost torque above this point. In other words, long runner performance can be enhanced by spark ignition timing aimed at lower rpm (more advance) than shorter ones (more retard). Ignition systems that allow this sort of selective cylinder timing are available.

It's also possible to juggle rocker-arm ratios to vary valve timing, if you don't have a camshaft that does this for you. In fact, the combination of multiple-lobe design cams and matched rocker-arm ratios is a popular topic of discussion these days. Overall, this tends to accomplish a torque curve somewhat diminished in peak value, but broadened over a wider span of rpm. Both off-the-corner and passing power can benefit at a somewhat reduced level of peak torque. You might want to give this some thought.

SOME MANIFOLD SELECTION AND USE GUIDELINES

In light of all the issues discussed in this article, here are some suggestions that are intended to help in the selection of intake manifolds.

1. If racing rules permit replacing the stock intake manifold, select an alternative that is suited to the rpm range that includes the lowest rpm expected on the track and the highest rpm expected for peak power.

2. Because of its inherently better cylinder-to-cylinder distribution characteristics, use a single-plane design rather than a two-plane—rules permitting.

3. If rules require the use of a two-plane manifold, include some sort of carburetor spacer to help alleviate air/fuel separation against the upper plane's floor at high rpm. Expect to make jetting changes that increase fuel flow.

4. While port-matching intake manifolds to head ports is important for high-rpm power, it can also enhance torque in the low- and mid-rpm ranges (provided that intake ports are smaller than those in the head).

5. Blocking exhaust crossover ports (either one or both sides) may improve mixture density from

reduced manifold heat, but more importantly, this prevents combustion contamination among cylinders that are fed exhaust gas during overlap periods of the connected cylinders.

6. Rough interior manifold surfaces may cause slight reductions in net airflow but (in wet-flow conditions) can improve mixture quality and combustion efficiency at the same time.

7. Short versus long runners in an intake manifold provide the opportunity for individual cylinder spark ignition and valve timing (see discussion of this issue earlier in this article).

8. Don't assume that cylinder-to-cylinder mixture distribution resolved on an engine dyno (or chassis dyno) will correlate directly to the racetrack. Vehicle dynamics and separated fuel randomly traversing a manifold's interior can combine to produce on-track distribution dissimilar to that observed on a dyno.

9. Plenum dividers (even portions of one) used in both single- and two-plane manifolds are crutches for increasing throttle response, along with low- and mid-rpm torque. Expect fuel-metering signal strength to vary according to the amount of divider used: Increasing divider material pulls more fuel, and decreasing it does the opposite.

10. Especially in single-plane manifolds, cylinder-to-cylinder mixture distribution can be regulated by the use of dams (leaner) or slots (richer) in the floor of the plenum. In these types of manifolds, you can also control distribution by altering pressure profiles at runner entries.

SOME AFTERTHOUGHTS

It is comparatively easy to overlook the impact of an improperly selected (or modified) intake manifold. The sophistication of today's racing engines requires careful thinking in the entire induction system process. Be honest about the rpm range in which the engine will not only operate but also require optimum torque. That is where proper manifold dimensioning is applied.

This chapter points out some of the not-too-often-published aspects of manifold design and function. It also includes study points that may require repeated reading to clarify new or different information. Based on our experience, it will likely be time well spent.

Chapter 14

OUTTAKES ON INTAKES

How to Shape Torque Curves Around Spec Manifolds

by Jim McFarland

Intake manifold (PN 2975)—Edelbrock's "bread-and-butter" Victor Jr. is capable of addressing a variety of circle-track engine needs. These include good cylinder-to-cylinder flow distribution, a low profile (if needed), responsive to one and two-inch carburetor spacers and works in a range of 5,000–8,500rpm. For rules not allowing spacers, a one-inch taller version (PN 2999) is available. Courtesy Edelbrock.

In time-honored fashion, racers maintain a constant vigil on rules interpretation and loopholes. As Junior Johnson once commented, "When they're looking at the ass end, you work on the front end. And when they're looking at the front end, you work on the ass end." Chicken farmers everywhere, beware.

But in the interest of reducing or maintaining costs of racing, certain engine components have come under the rule of "no modification." Intake manifolds often fall into the "spec" category, placing responsibility on the manufacturer to optimize engine performance by redesign through identification and correction of defects in the stock parts replaced.

This chapter carves out some of the issues that enable engine performance optimization, confined to spec intake manifolds, by examining certain companion parts that can be tailored to "regulated" airflow. It would appear that rules evolve at all levels of racing, including those promulgated by individual tracks. It's not an objective of this chapter to draw lines on how such decisions are made. Rather, we'll just get into ways engine performance can be boosted by attending parts selection. But first, some basics about intake manifolds provided in the interest of building an informational foundation for further discussion.

BASIC INTAKE MANIFOLD COMPONENTS AND FEATURES

Since the vast majority of current circle-track engines use carburetors, we'll confine this material to carbureted or "wet flow" intake manifolds. As such, these comprise a network of passages (runners) connecting inlet ports to a space (plenum). Runners can be connected to the plenum in such a way that separates every other cylinder in the firing order or terminates in a common plenum irrespective of firing order.

In the case of the former, more than one level or "plane" is often used, allowing cylinder separation so intake charges are delivered alternatively from one plane to the other. Typically, these are labeled "two-plane" in design (common to V-type engines), while manifolds with runners connected to a common plenum are designated "single-plane."

Historically, the range of engine rpm in which the greatest volumetric efficiency (torque production) is achieved characterizes the differences between two-plane and single-plane intake manifolds. Despite efforts by some manufacturers to shrink single-plane manifolds to perform similarly to two-plane (and the opposite), two-plane manifolds remain the choice for low- and mid-rpm torque, while single-planes are most efficient at higher engine speeds. Generally, these are the fundamental differences between the two types.

Of the features that affect intake manifold "tuning," runner cross section and length combined with plenum volume can be dimensionally united to improve torque and throttle response. Generally, runner cross section affects the rpm point at which peak torque is produced, while length influences the torque produced above and below this point. Plenum volume, where sufficient space is available, can be made to behave much like a header collector, boosting torque below peak torque rpm. However, in most instances plenums are configured to assist in the transition of air/fuel mixtures from carburetor to manifold runners.

Are other features of an intake manifold important? Certainly. One is the relationship between the carburetor base and plenum floor. Visualize air and fuel being discharged from beneath the carburetor's throats. Two issues are immediately critical: the dynamics of "turning" air/fuel mixture into intake manifold runners (without encountering air/fuel separation) and the proximity of the carburetor base and plenum floor. Through the use of carburetor spacers, both of these problems can be addressed successfully, if rules permit their use.

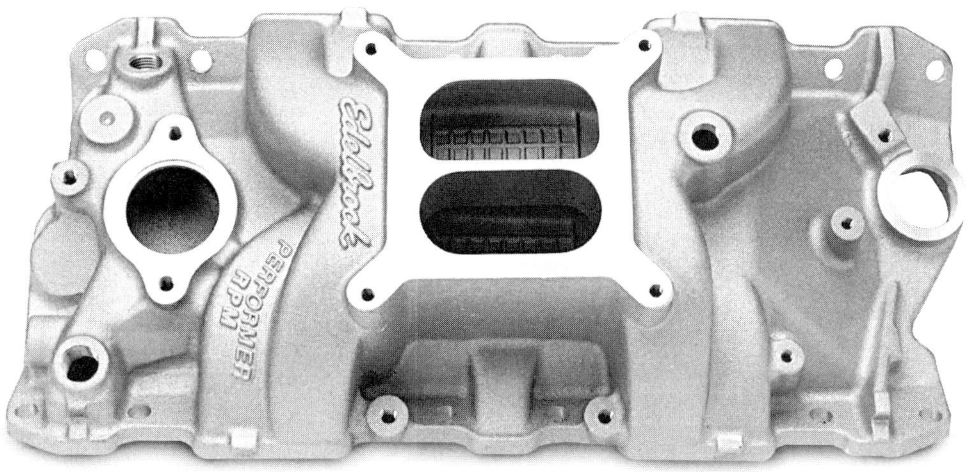

Intake manifold (PN 7101)—Designed to operate best in the 1,500–6,500 rpm range, the Edelbrock "Performer RPM" is intended for small-block Chevys displacing 262–400 cid, produces a wide torque band, includes throttle response features of a two-plane design and has runners sized for mid- and higher-rpm power. Courtesy Edelbrock.

Spacers provide an opportunity for air/fuel mixtures to lose a measure of kinetic energy, leaving a carburetor and entering manifold runners. Should you not feel this issue is critical, consider the possibility of mixtures moving in excess of 200 feet/second and changing direction from vertical to something approaching horizontal . . . over a carburetor base-to-plenum floor distance of less than six inches. The effect is akin to sticking a water hose in a bucket. Increasing carburetor height helps alleviate the problem of disrupted mixture quality during high rpm, as carburetor size decreases, or both.

WHY STOCK AND PERFORMANCE MANIFOLDS DIFFER

It's not just about power. Aside from providing designs intended to increase airflow, improving cylinder-to-cylinder and air/fuel mixture delivery is important to increased engine parts durability. In stock intake manifolds, it is not uncommon to discover cylinder-to-cylinder air/fuel differences upward to three or four ratios, from the leanest to richest cylinders, traceable to improper manifold design. For this reason alone, particularly when stock intake manifolds are used for racing, mixture inequality can lead to damaged engine parts (pistons, rings and cylinder heads).

If racers were permitted to make manifold modifications that improved cylinder-to-cylinder air and/or mixture distribution, problems of this nature might be resolvable. However, since improved distribution is among the design criteria of aftermarket manifold designers and the modification of stock manifolds can be expensive, engine durability is often a benefactor from the use of spec intake manifolds. The fact they produce additional power is a racer plus.

RELATING MANIFOLD SIZE TO ENGINE SIZE AND RPM

By way of quick review, barring any changes to the internals of an intake manifold, the larger the engine or higher the rpm, the lower the engine speed at which peak manifold efficiency is achieved and here's why.

There is an airflow velocity (sometimes called "mean flow" velocity) observed at or near peak torque, dependent upon piston displacement, rpm and flow-path cross section. Although this dimension is fixed in a spec manifold, knowing its influence upon torque production in the engine you're using is helpful.

Of the methods that allow you to pinpoint engine speed range where a specific intake manifold's runner cross-section area tends to boost torque, the following one should be familiar after reading Chapter 13. The arithmetic goes as follows:

Measure and calculate the area of each intake runner entry and exit (at the cylinder head flange). Add the two values and divide by two, producing an "average" runner cross-section area. Now determine the volume of one cylinder (displacement divided by the number of cylinders). Armed with this information, use the following equation: Torque peak rpm = (88,200 x average runner cross section) divided by the volume of one cylinder. If we assume and plug in some numbers, the results look like this:

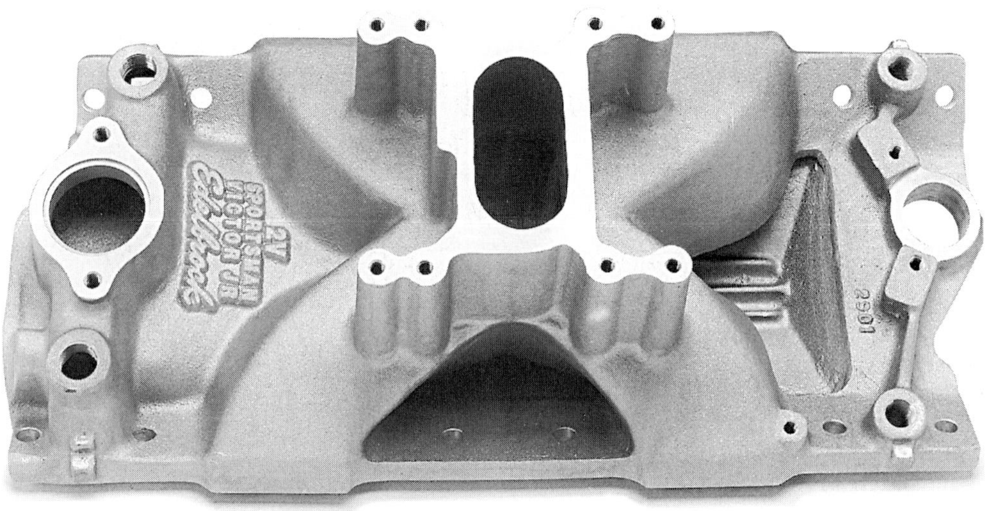

Intake manifold (PN 2901)—Addressing the 2V Sportsman engine market, Edelbrock's Victor Jr. Sportsman 2V features a raised carburetor flange (helping to eliminate the need for spacers or when none are allowed), a 3,500–7,000rpm operating range, is port-sized for GM Vortec cylinder heads. Courtesy Edelbrock.

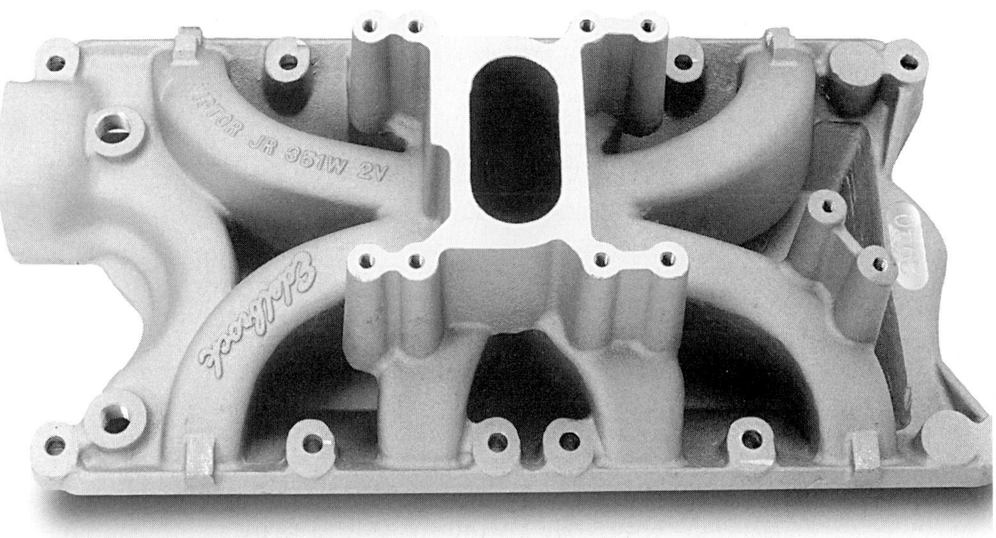

Intake manifold (PN 2940)—Edelbrock's Victor Jr. Ford 2V Sportsman manifold is a "spec" design available for 9.2-inch blocks. The PN 2941 is its counterpart that fits blocks of 9.5-inch deck height. Range of best operation is 3,500–7,000. Manifolds are intended for use on cast-iron SVO, Dart or Edelbrock Victor Jr. cylinder heads. Courtesy Edelbrock.

Peak torque rpm = (88,200 x 2.9 sq. in.) + 43.75ci = 5,946rpm

Where 43.75ci is the cylinder volume of a 350 V-8 engine, 2.9 sq. in. is the average cross section of an intake manifold runner, and 88,200 is a constant used for units and related conversion factors.

Why is this exercise of value? You now know the rpm at which this particular spec intake manifold is designed to boost volumetric efficiency (torque) and the engine speed around which you can begin optimizing other components to support the manifold's inherent performance range. This is particularly valuable because parts companion to the intake manifold can now be selected (or modified) to contemplate where the manifold is not working best and integrating their performances accordingly.

For example, if you're dealing with an intake and it's boosting torque in an rpm range where you're not running the engine (but should), some gearing or tire size changes may be in order. You may also want to consider some dimensional changes to the exhaust system. Perhaps investigate a different camshaft or where the one you have should be positioned (advanced or retarded). In short, you need to begin building torque in a range of engine speed where the "fixed" intake is not.

On the other hand, even if the intake manifold is working best in the span of rpm being targeted, changing or modifying companion parts to further boost torque in this range could be of additional help. But in either case, knowing where the manifold is providing the greatest gains in power can help you analyze the overall engine package and its performance.

In particular, this information can lead to more sensible selections of other engine parts as they relate to the rpm range in which you plan to run the engine. In a sense, dealing with spec intake manifolds is not much different than the use of cylinder heads for which no modifications are allowed. You need to make decisions about (1) a specific span of engine speed where power is required, (2) performance limitations of the parts for which no alterations can be made (in this case intake manifolds) and (3) characteristics of companion parts that can help compensate for rules restricting modifications.

GENERAL GUIDELINES FOR COMPANION PARTS SELECTION

Let's begin this with an example case. Suppose we have an engine of 355ci displacement (44.375 ci/cylinder). The "spec" intake manifold is a single-plane 4V design. By measuring its runner entry area (2.75 sq. in.) and exit area (1.98 sq. in.), we determine an average of 2.37 sq. in. Using the previously supplied equation, the peak torque point (for this intake manifold and assuming all runners of equal cross section) computes to 4,710 rpm.

Intake manifold (PN 300-110)—According to information from Holley, this "spec" intake manifold design (300–110) features tall runners exiting the plenum and increased runner "twist" near cylinder heads, intended to improve cylinder-to-cylinder mixture distribution. Additional design features are included that help provide a broad and flat torque curve for improved on-track drivability (torque for off-the-corner and passing situations). Courtesy Holley.

Now, we need to examine the rpm range in which we'd planned to run the engine. Perhaps we'd previously thought something like 4,500–7,000 was acceptable, but now we discover the "spec" manifold will be on the declining side of its torque production on the low end of our projected rpm range. Either we adjust the intended rpm range downward or pick companion pieces that boost torque above the manifold's range of efficiency. In short, we'd like to make the manifold "look bigger" to the engine.

In fact, we might even consider where this manifold is working (relative to rpm) as a plus. The engine can be fitted with a cam and set of headers that work best in the 5,000–7,000 rpm span, leaving it to the intake manifold to assist torque production for off-the-corner acceleration. On the other hand, we'd be fooling ourselves by placing dependence on the manifold to produce much useable torque if the engine was spending most of its time above 5,500 rpm. Among the reasons for identifying rpm at which an intake manifold is best suited, this is clearly one.

On the chance you might not have yet selected a set of header dimensions, suppose we examine that next. With the knowledge that primary pipe cross-section area influences an engine's peak torque rpm point (regarding contributions by headers), we decide a torque boost at 5,500 would probably be a sensible place to start considering on-track passing requirements upon exiting a corner.

By performing a little manipulation of the equation provided earlier:

Cross-section area = Peak torque rpm x Cylinder volume + 88,200

If we plug in the values for our example engine, primary header pipe o.d. needs to be 1.94 inches, in order to produce a torque boost at 5,500. Since 2.0 inches of primary pipe o.d. is the nearest standard size to 1.94 inches, a set of headers built with pipe of 2.0 inches o.d. should get the job done.

But maybe you already have a set of 1.750-inch pipes and need to determine how they match up to the "spec" manifold we're discussing. If primary pipe wall thickness is 0.040-inch (not uncommon), o.d. is 1.67 inches and the area 2.19 sq. in. Using the same mathematical approach applied to our "spec" intake manifold, peak torque rpm for our headers is about 4,350 rpm. Now we have headers peaking their efficiency in the range of our spec manifold.

One solution to this problem is to shorten primary pipe length, thereby adding torque above the peak and removing it from below. However, this could be considered a trade-off to properly addressing the situation. Using a set built with 2.0 inches of pipe o.d. (as outlined above) not only goes

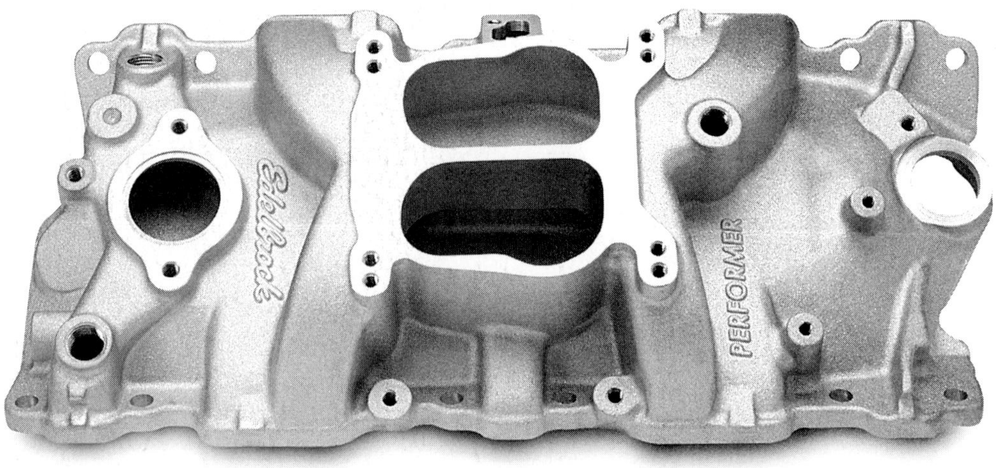

This small-block Chevrolet spec manifold from Edelbrock shows an intent to equalize runner lengths, as compared to a stock manifold, although some length variation lends itself to individual cylinder tuning. Rocker-arm ratio and primary header pipe are two possible "tuning tools" for "spec" manifolds with greater runner length variations than this design. Courtesy Edelbrock.

While particularly beneficial when using two-plane "spec" manifolds, two-inch spacers (such as this "Swirl-Torque" design from TD Performance Products featuring mixture-enhancing slots) can aid cylinder-to-cylinder mixture quality and distribution during mid- and high-rpm operation. Power gains stem from increased carburetor-to-plenum floor height (on the upper plane), especially with the two-plane design, although similar improvement can be found with spacers on single-plane manifolds. Courtesy TD Performance Products.

directly to the issue but also provides some additional torque-shifting opportunity when shortening or lengthening this size pipe.

In the camshaft department, it's a general rule of thumb that you specify an rpm range in which you'd like the most significant power gains. Since you're already in the "rpm determination" mode, sharing with your camshaft grinder the ranges of engine speed you'd like to address is not only valuable but also fairly available information. Plus, once you've decided upon a camshaft suited to filling in blanks not provided by the spec intake manifold, advancing and retarding its installed position can further aid the cause.

SOME CONCLUDING COMMENTS

Conforming companion parts to a spec intake manifold involves changing the basic shape of an engine's torque curve. In particular, header design and dimensions and valve timing patterns can be used favorably. Generally, the first order of business is to determine the rpm range in which the spec manifold is most productive.

By comparing this information to the span of rpm in which the engine will be used (or should be used), other parts can be selected to augment areas not supported by the "spec" manifold. While cylinder-to-cylinder air and mixture distribution improvements provided by spec manifolds benefit engine parts durability, it's important to determine where power is needed and the manifold is not supportive.

Spec Manifold Limitations

Generally, you'll find that unless a spec intake manifold has been designed to address the problems of an otherwise stock version (or the subject is a racing manifold with certain deficiencies), there are some problems you need to consider. Of them, cylinder-to-cylinder air or air/fuel mixture distribution (or both) is common. Aside from the fact this issue is fundamental with the design of any intake manifold, some finished products are worse than others. Intake manifolds designed primarily to increase net airflow in deference to improved distribution are cases in point.

Stated another way, intake manifolds that deliver differing amounts of air or varying qualities of air or air/fuel mixtures tend to prevent all cylinders from developing equal or near-equal amounts of power. Intake manifolds with poor cylinder-to-cylinder delivery of the "working fluid" create opportunities to compensate for this problem.

Several of these opportunities will be addressed in later paragraphs.

The point here is that treating all cylinders on the assumption each will produce equal levels of power is flawed by spec intake manifolds that create distribution problems. Therefore, by identifying the "problem" cylinders and making adjustments that compensate for (or correct) these conditions, it's possible to improve the overall power of an engine package. Keep that thought foremost in your mind as we dig into the details of how this can be done.

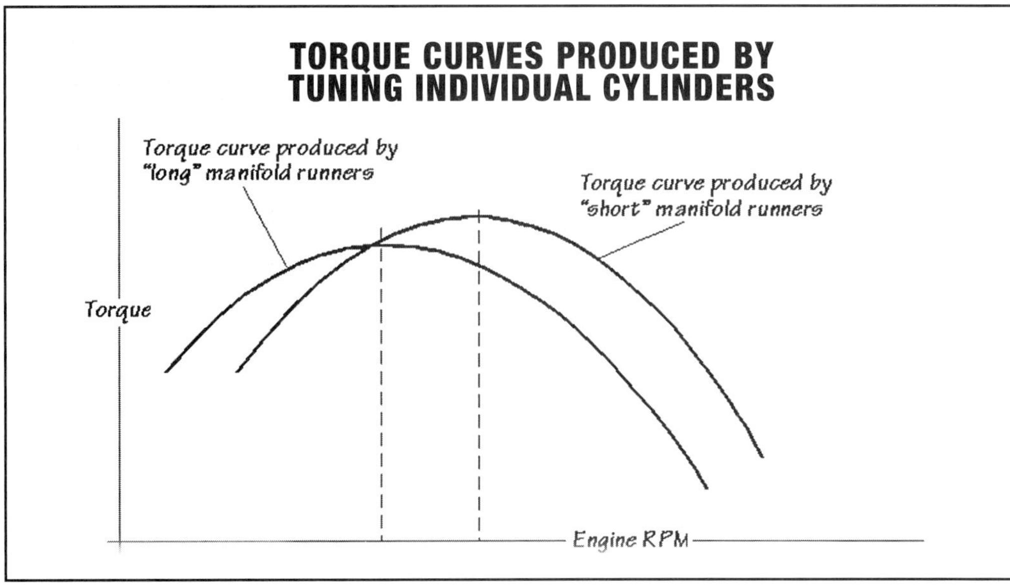

Although somewhat exaggerated in this illustration, the effects of tuning short and long "spec" intake manifold runners tends to produce a wide, flat torque curve . . . the effects of which are the result of two "independent" torque curves or a composite curve that broadens the effective range of an engine by way of optimizing the output of cylinders served by runners of different length dimensions. Other factors influence this type of engine output, as discussed in the accompanying story. This tuning process actually produces a "family" of torque curves, the net of which can extend the range of operation . . . particularly beneficial in many circle-track situations.

Begin with an Airflow Study

If possible, do an airflow study of the spec intake manifold, bolted onto the cylinder head(s) to be used. Flow each runner/port combination by itself, sealing off all other passages when you do. In this fashion, you'll develop an idea about the airflow potential of each runner/port combination, leading to some indication of cylinder-to-cylinder air distribution. While this method does not contemplate certain "dynamic" conditions (so-called "cross talk" or pressure excursions) created in a running engine, it still provides workable indications.

Perform these tests throughout the range of valve lifts that accompany the camshaft you'll be using. Pay particular attention to airflow at valve lifts of about 65 percent of net for the intake and 85 percent for the exhaust. Each of these two percentages represents piston positions at which "critical" intake and exhaust flow occurs during engine operation. These are also points of valve lift where piston position and port flow are important (during engine operation) for optimizing volumetric efficiency (intake stroke) and cylinder blow-down (exhaust stroke).

If the manifold has problems, find out where they are and which cylinder head intake ports and manifold runners are sources of reduced or quality-impaired flow and potential mixture-quality disturbance. This is basic and required information, without which you're operating in the dark regarding tuning "one cylinder at a time." Tuning adjustments to be discussed later cannot be properly directed to specific cylinders without some identification of "good" and "bad" runner/port combinations.

TUNING ONE CYLINDER AT A TIME

Once identification of "good" and "bad" intake paths is made (either by airflow study, spark-plug readings, air/fuel ratio measurement or a comparable method), companion parts or "tuning methods" can be employed to optimize individual cylinder output. An initial objective is to determine which engine components or functions can affect individual cylinder output. With this information in hand, you can begin making adjustments for improved performance of the entire engine package.

Following are some areas of typical investigation. For further validation of the importance placed on areas of potential power gains, two prominent engine builders were consulted on the spec manifold issue: Keith Dorton, owner of Automotive Specialties Racing Engines, and Dennis Wells, proprietor of Wells Racing Engines. Both have extensive engine building experience that includes optimizing power from "spec" intake manifolds.

We requested two sets of responses from these professionals. One was to give priority to a list of engine components or systems that can influence spec intake manifold performance. The other involved a list of four questions, each directed to

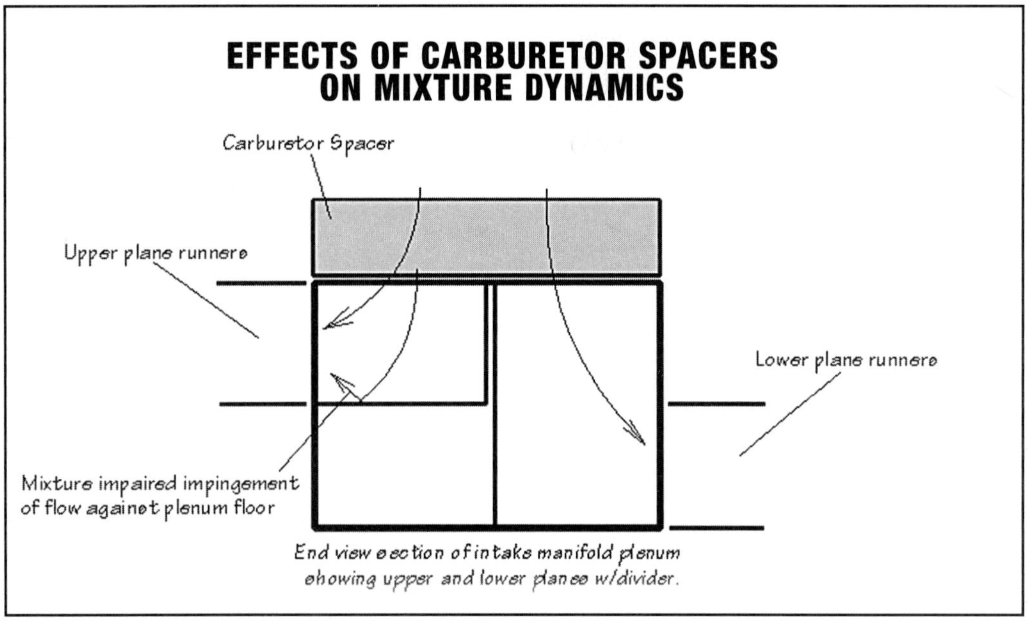

Particularly in two-plane spec manifold designs, carburetor spacers where permitted can provide additional room for exiting air/fuel mixtures to lose energy and negotiate turns into upper plane runners. This aides mixture quality and, depending upon circumstances, benefits overall combustion efficiency and power. Impingement of mixtures on the plenum floor can further degrade cylinder-to-cylinder mixture distribution.

dig further into the experience base these two builders have amassed.

Following is the list of components or systems proposed for placement in order of importance and tendency of an engine to be sensitive to change:

• Air cleaner design. This category includes size of filter element and design of cleaner assembly lid and base.
• Carburetor spacer. Here we refer to spacer use, height and design (open, four-hole or hole configuration).
• Header system. Includes primary pipe steps, pipe size, collector dimensions and use of cross-over pipes.
• Camshaft design. Not excluded was the possibility of multiple intake/exhaust lobe profile specifications and positioning on the shaft.
• Ignition timing. Sensitivity of a spec manifold engine to variations in spark timing.
• Carburetor. Not restricted to size alone, this category includes booster design and other possible mixture distribution "fixes."
• Rocker-arm ratios. In particular, the issue here is potential benefit of varying ratios among cylinders.
• Spark plugs. This deals with heat ranges, indexing and depth of plug into the combustion chamber.

Here is the order suggested by Dennis Wells:

• Air cleaner design
• Header configuration
• Carburetor spacer
• Rocker-arm ratio
• Camshaft specifications
• Carburetor design/modification
• Ignition timing
• Spark plugs

Here is the order suggested by Keith Dorton:

• Rocker-arm ratio
• Ignition timing
• Carburetor design/modification
• Camshaft specifications
• Spark plugs
• Carburetor spacer
• Header configuration
• Air cleaner design

To expand on these topics, consider the following additional thoughts about some of the topics and priorities listed.

Air Cleaner Design—While the amount of filter surface area and top configuration of a given air cleaner assembly can affect net airflow, the shape of an assembly's bottom is often more critical. Since air tends to "take the shortest path" in an induction system, the distance from the air cleaner base to the carburetor entry frequently represents this path.

Therefore, the manner in which air moves across the floor of the base and enters the carburetor can affect net airflow. Experimenting with base shapes often helps compensate for "spec" manifold runners otherwise experiencing flow problems. It's a way for helping a bad condition become somewhat better.

Header Configuration—Sizing primary pipes to match "good" and "bad" intake manifold runners can be an effective way of optimizing individual cylinder power. For example, a port deficient in flow may be "tuned" for a lower range of rpm by lengthening its companion primary header pipe. Ports flowing higher levels of air may be tuned to higher rpm by shortening pipe length. In a similar fashion, pipe diameter can also be used as a tuning tool (low rpm vs. high rpm).

Since header collectors tend to be torque-productive at engine speeds near or below peak torque rpm, general improvements in torque within this range can further support "spec" manifold runners limited to lower rpm. Generally, the greater the collector volume, the higher their output below peak torque rpm.

Rocker-Arm Ratio—This can be a very effective way of aiding poor-flowing intake manifold runners. Slight increases in rocker ratio can boost volumetric efficiency, leading to improved net port flow for runners incapable of matching the flow provided by other and more efficient runners. In such cases, increasing the ratio on the intake side is often the most effective area in which to work.

Camshaft Design—Depending upon how friendly you may be with your favorite camshaft grinder, this is an area of potential benefit. The reduced airflow of certain "spec" manifolds runners can be addressed by increasing the specifications of intake lobes (cylinder by cylinder), increasing valve overlap periods for the same cylinders and (in some cases) incorporating higher lift values can help equalize or balance cylinder-to-cylinder airflow. Delaying the intake closing point is another area of sensitivity.

In short, each intake runner's flow limitations can be addressed by tailoring intake and exhaust lobe specifications (including displacement angle) to match the flow deficiencies. The result, at least in this writer's experience, can be a finished camshaft that sports a variety of intake and exhaust lobes, frequently positioned at differing displacement angles along the shaft. You might want to give it some thought.

Ignition Systems—Where allowed, the use of "programmable" ignition systems is providing individual cylinder spark timing ... or at least a formatting of spark maps that are more applicable to optimizing single-cylinder tuning than systems that don't. As this technology continues to evolve, racers will likely have access to increased ability for addressing the specific spark requirements of cylinders that vary in volumetric efficiency, one from another...for whatever the reason.

Summary of the engine builder interview. The following four questions were asked of each builder. Their responses are noted accordingly.

Question #1: Based on the list of components and systems listed, which do you feel can be the most influential for adjusting for spec manifold performance, although not necessarily the easiest change?

Dorton: "From what we've seen, adjusting valve timing (primarily through the selection of rocker ratios) and establishing individual cylinder spark timing appear to be the two most effective areas for optimizing overall power."

Wells: "Header configuration seems to have the most effect on power output when trying to adjust for low-performing cylinders. By matching primary pipe length to variations in individual manifold runner airflow, we've found it's possible to boost torque in an rpm range where the cylinder seems to perform best. Of course, primary pipe diameter is also a tool that helps move torque around in the rpm range

PORT FLOW VARIATIONS FOR SINGLE CYLINDER TUNING METHODS

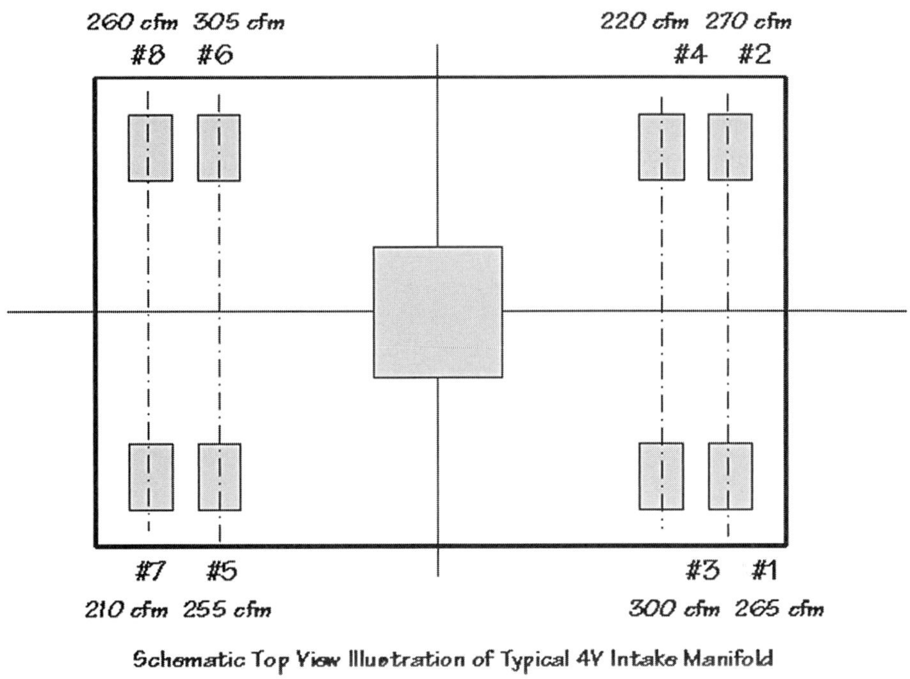

This sample arrangement of intake path flow volume (cfm) is a theoretical representation of variations that may occur in a "spec" intake manifold . . . perhaps somewhat exaggerated to illustrate a basis for companion parts selection.

Question #2: Some pretty good results have been produced tailoring camshaft lobes and displacement angles to specific cylinders. This seems to work for "spec" manifolds with differing length of runners or airflow variations runner to runner. Any thoughts on this?

Dorton: "I'm sure there are benefits to be derived

by this method. However, my opinion is that it requires an extensive amount of testing to arrive at the best combinations of lobe specifications."

Wells: "This approach has shown to be a good way to help balance out the differences that may occur in "spec" intake manifold runner lengths. In other words, the longer the runner the lower in rpm it tends to make torque. The shorter the runner, the higher the rpm where it tends to make torque. Camshaft lobes can be designed to address these conditions."

Question #3: Pertaining to the above question, there have been header systems sized and built to directly match "spec" intake manifold runner lengths. The results tended to produce engines with flatter torque curves than those using headers that didn't incorporate this feature. What can you say about this approach?

Dorton: "We have used many different header primary designs, including step, tri-Y and expansion chambers, with sometimes huge differences in power. Although I'm not sure about the effects of these designs on individual cylinder power, I expect there are gains to be made by experimentation."

Wells: "I think this will work well but only so long as you keep the car geared to take advantage of the additional torque that's available when coming out of the corners. By this I mean that if you've built additional torque into the lower or mid-rpm range and don't gear the car to run in this range, your efforts will be lost."

Question #4: Depending upon the airflow limitations of a given "spec" intake manifold, do you think it's best to determine the rpm range where the manifold works best and then select engine parts around this range. If so, how would you go about finding out what this rpm range is without running the engine?

Dorton: "Things such as carburetor size limitations and camshaft restrictions will determine the engine's rpm range. In many cases, rules limit what can be done to an engine, thereby causing airflow limits. Even in these cases, optimizing individual cylinder airflow, within the range of restricted rpm, is a worthwhile path to take."

Wells: "This is a good approach. It can save a lot of time and reduce mistakes made in choosing engine parts. What I have done, and do, is use one of the engine simulation programs currently available. A little personal computer time spent with one of these programs can really be worthwhile. Specific power levels predicted may not conform to actual test data, but you can certainly see trends and eliminate many of the variables that can lead to buying the wrong parts."

Some Concluding Thoughts

Single-cylinder tuning is becoming more common in the building of contemporary race engines. Even the OEMs have gotten into this act, targeting increased fuel economy and emissions reduction in the process while aided by the use of on-board electronic fuel and spark management systems. However, even in the absence of electronics, there are mechanical ways of accomplishing similar "individual cylinder power" objectives.

But as suggested at the outset of this chapter, it's important to identify which runners in a "spec" intake manifold may be flow deficient, of "unequal" length or both. And when you can, do the "airflow math." Even armed with this information, it can be a task to deal with problems these conditions create. Without such information, however, the job becomes more difficult.

For the best results, analysis becomes a combination of several data inputs, some derived before an engine is run and some afterward. Most important, recognize the fact an engine is a system of individual cylinders, each of which must be optimized (within the range of most frequent engine operation), if a maximum power package is to be assembled. In this particular case, the whole is the sum of its parts.

Chapter 15

CARB BASICS

Getting the Most From Your Mixing Valve

Text and Illustrations by Jim McFarland

Of the objectives carburetors need to accomplish, the most important one is providing properly proportioned fuel that enables correct air/fuel mixture ratios throughout a wide range of engine speed and load. Installed on an engine dynamometer test stand, meeting these goals is far less difficult than in the hostile environment of a race car on the track.

The purpose of this chapter is first to lay out the basics of a carburetor's functions, essentially to form a basis of understanding when we get into discussing performance-specific adjustments and modifications.

HOW A CARBURETOR WORKS

First of all, fuel is not "sucked" into an engine. It is forced in by atmospheric pressure. More specifically, a pressure differential is created between the force of atmospheric pressure and cylinder pressure, which is typically lower. Keep in mind there are times, particularly during reversion conditions (early period of an intake cycle), when cylinder pressure may be higher than atmospheric. But since lower-than-atmospheric cylinder pressure is the predominant condition, atmospheric pressure forces air into an engine.

OK, that's thc dominant source of airflow. Airflow rate is a function of several variables. These include total piston displacement, engine rpm, size of intake passage and carburetor openings. Within a carburetor throat, there is a reduction in passage diameter called a "venturi." This is placed so as to increase airflow rate in an area exposed to fuel delivery paths and reduce pressure in this region (see illustration on next page). Under the influence of atmospheric pressure, an assortment of fuel-delivery passages (idle, idle transfer, intermediate, and main) combine to provide fuel in quantities proportional to airflow.

All this, of course, is in a perfect world, which we know doesn't exist. In reality, pressure conditions around a carburetor's entry are not always equal to atmospheric. Depending on such factors as air-cleaner design, under-hood pressure and/or temperature conditions and other air delivery variables, real-world carburetors don't always experience atmospheric entry pressure. This is one reason that optimized carburetor calibrations on an engine dynamometer don't always correlate to on-track calibration requirements.

Bottom line, carburetors are "pressure differential" devices that deliver fuel on the basis of low-pressure conditions downstream of the fuel delivery point. Multiple circuits are designed to provide metered fuel as a function of engine speed, load, volumetric efficiency and power level. For best and most consistent performance, carburetors need to be located in environments that attempt to provide near-atmospheric pressure conditions.

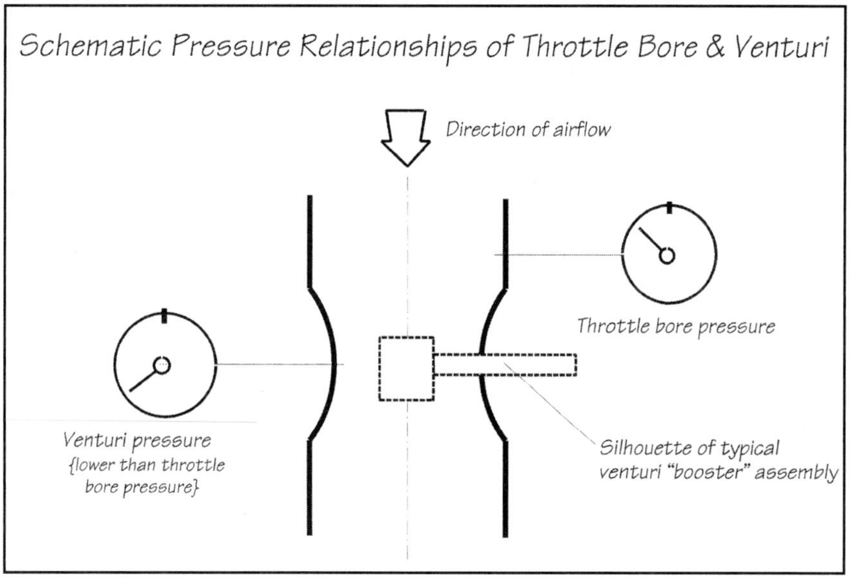

Constriction (a venturi) placed in a carburetor's throttle passage causes an increase in flow velocity, thereby reducing flow in this region as compared to the throttle bore or atmospheric pressure. This increases the amount of fuel delivery (metering) signal experienced by fuel in the fuel bowl. Booster design and placement are critical to metering signal strength and fuel atomization efficiency.

When the throttle is opened quickly, airflow can exceed fuel-delivery capacity of other carburetor circuits. To maintain proper air/fuel mixture ratio during these conditions, an accelerator pump circuit provides fuel supplied from the fuel bowl, delivered under pressure by an accelerator pump and discharged through nozzles of specific size according to fuel quantity required.

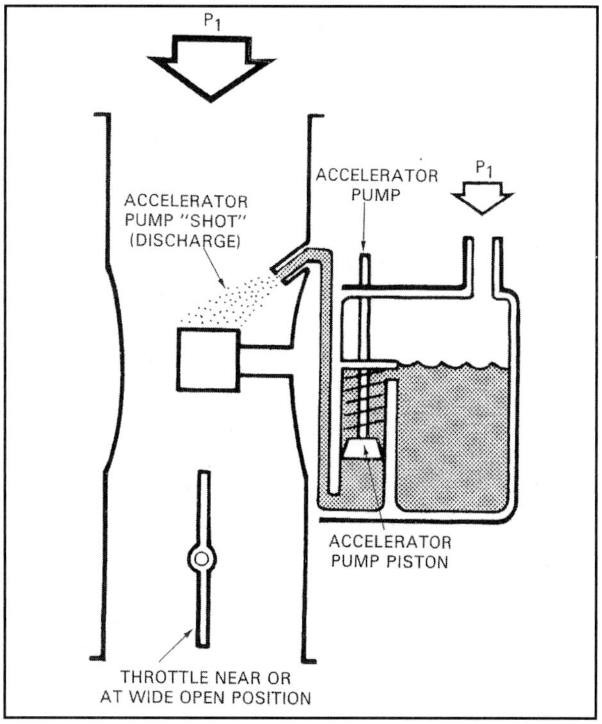

BASIC CARBURETOR CIRCUITS

In the illustrations provided, fundamental fuel delivery circuits of a carburetor are shown in schematic form. Expanded captions for each are supplied to help in understanding the circuits shown.

It is important to understand each basic circuit tends to "overlap" the previous one. For example, once the idle circuit becomes "saturated" (reaches its limit of fuel delivery), it continues to provide fuel at that level throughout the remainder of the rpm range. It simply makes a contribution to the overall quantity of fuel delivered.

The accelerator pump circuit is an exception to this since it only delivers a volume consistent with the capacity of its pump reservoir, size of discharge nozzles and duration of delivery. It's also a "cover up" for the brief time required before main system fuel flow is activated.

Another area of sensitivity, particularly regarding fuel atomization efficiency, pertains to venturi booster design and placement. It is here liquid fuel, delivered to the booster's venturi, is "sheared" or "mixed" into droplet size. Keep in mind, the time required for fuel to burn or be combusted includes the effect of its size. The larger the droplet the longer the time required to burn. Air/fuel mixtures comprised principally of large droplets tend to act "rich," thereby exhibiting slower burn rates than mixtures of smaller size.

As part of the main fuel delivery circuit, the manner and efficiency in which boosters atomize fuel is critical to downstream air/fuel mixture quality. The smaller the atomized particles, the more they "behave" like air. This causes them to follow circuitous intake manifold and intake port paths with less mechanical separation from air than larger droplets.

The emphasis here is that attention should be given to making sure boosters are firmly anchored in the carburetor's main body. Loose boosters tend to draw air outside the venturi, causing disruption in atomization efficiency and potentially lost power. It's also important to investigate booster designs and claims of manufacturers, placing emphasis on finding out which provides the best atomization and why. If you run into a manufacturer who cannot (or will not) share this type of information, it may be time to select another source.

MATCHING CARBURETORS TO PISTON DISPLACEMENT AND RPM

If we define an engine's air consumption ability as a function of volumetric efficiency (a percentage comparison of air ingested vs. total piston displacement), matching carburetor flow capacity to volumetric efficiency is a required objective in achieving optimum on-track engine performance.

Several time-honored methods can be used to match carburetor flow capacity to engine size and rpm. Of them, the following is a frequent choice.

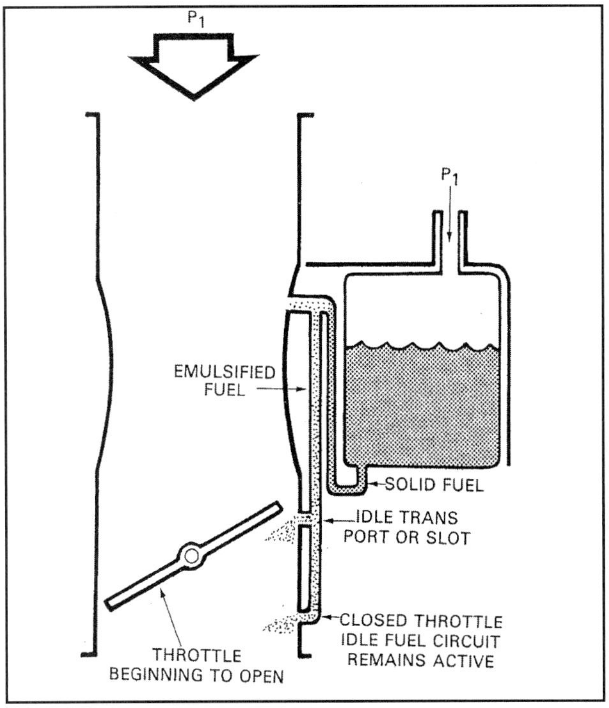

As additional throttle blade opening is provided and prior to activation of the booster venturi (main) circuit, the idle-transfer circuit supplies additional fuel. A port is exposed for fuel delivery, still utilizing the idle air bleed opening, enabling emulsified (mixed with air) fuel supply.

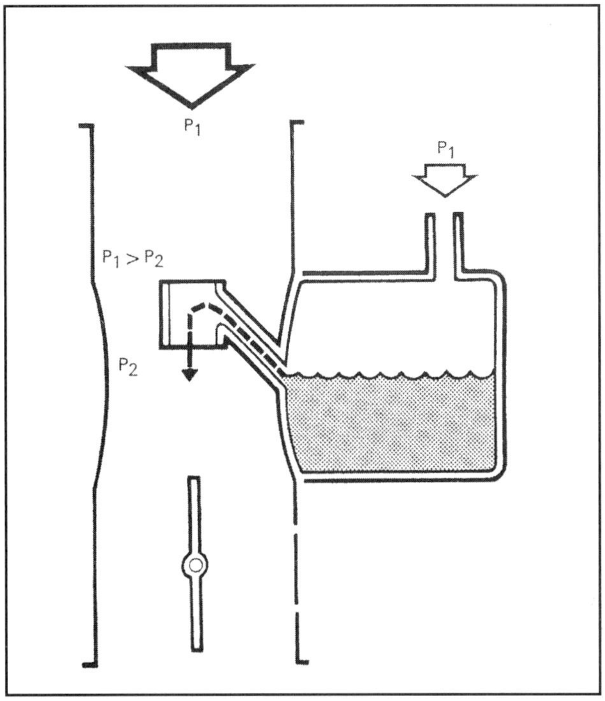

In a normally aspirated engine, unless upset by pressure around the carburetor entry, P_1 remains relatively constant and equal to atmospheric pressure. At the minimum venturi diameter, the difference between P_1 and P_2 is the greatest, causing the highest metering signal strength possible. P_1 at the fuel bowl vent to atmosphere is considered the reference pressure and should be maintained as close to atmospheric pressure as possible. Otherwise, fuel delivery through the main jets will be upset.

For example, divide total piston displacement (cubic inches) by 2. Divide the answer by 1,728 to convert cubic inches/revolution to cubic feet/revolution. Multiply this answer by the rpm at which you expect maximum carburetor airflow. Assume a volumetric efficiency percentage reasonable for the engine at the rpm you've decided is the maximum. Multiply this maximum rpm by the volumetric efficiency to obtain the cfm (cubic feet/minute) of carburetor required. Assuming an engine of 400 cid and plugging numbers into all these words, it looks like the following:

400 c.i.d. ÷ 2 = 200 c.i.d.
200 c.i.d. ÷ 1,728 = 0.116 cubic ft./revolution
0.116 x 7,000 = 810 cfm

Assume a volumetric efficiency of 90 percent (0.90) at 7,000 rpm.

810 cfm x 0.90 = 729.16 cfm

This final airflow number equates to the theoretical flow capacity of a carburetor sized for a 400-cid engine with a volumetric efficiency of 90 percent at 7,000 rpm. While this is a pretty good rule of thumb for basic carburetor sizing, instances do arise when a larger carburetor (in this same engine) might make more power. A common but often misunderstood reason is tied to mixture quality delivered by the larger carburetor, assuming both carburetors are positioned the same distance from the intake manifold's plenum floor.

Specifically, the larger carburetor tends to produce lower exit velocities of air/fuel mixtures, compared to the smaller carburetor. Because of this, mechanical separation of air and fuel (being impinged on the plenum floor) tends to be less with the larger carburetor. So, with the same volume of delivered air and fuel, the larger carburetor makes more power on the strength of improved mixture quality. You'll see more of this discussion in the subsequent section on carburetor spacers.

THE ROLE OF CARBURETOR SPACERS

Here is an area of carburetor application that is sometimes misunderstood. The majority of intake manifolds have flat or near-flat plenum floors. Air discharging from a carburetor's throats exhibits some level of kinetic energy. It's also compressible and more capable of negotiating directional change than fuel. Factors that affect the impingement of air/fuel mixtures on a plenum floor include carburetor size, increased rpm, and the distance from carburetor base to plenum floor.

While there are some beneficial effects from adding plenum volume, physical distance between floor and carburetor base appears more critical—especially at high rpm where mixtures can slam into plenum floors much like the effects of aiming a water hose into a bucket. Simply raising a carburetor 1–2 inches from the intake manifold's designed flange height can help alleviate the delivered-mixture problem.

Generally, the smaller the carburetor, the higher it needs to be off the plenum floor. Of course, this tends to weaken fuel-metering signals since the carburetor is being located further from the runner entries, the manifold's strongest source of signals. But if proper recalibration steps are taken, particularly to low rpm and off-idle fuel delivery, previously lost throttle response and torque can be restored.

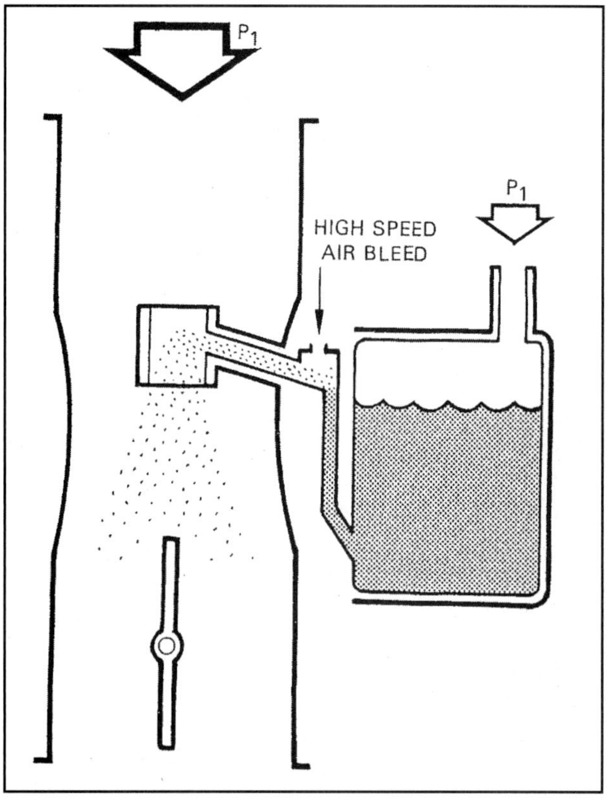

When engine rpm or load requires fuel delivery, which exceeds the low-speed fuel requirements, the "main" system is activated. Main system orifices (typically jets) control this circuit's fuel flow, emulsified by air entering the high-speed air bleeds, just prior to mixture discharge through the venturi boosters. Many experienced engine builders/tuners suggest modification of high-speed bleeds only if proper exhaust gas analysis is performed.

A FINAL WORD ON CARBURETORS AND COMBUSTION EFFICIENCY

Regardless of all the other issues discussed in this chapter, remember this: carburetors are intended to mix fuel with air, and the more efficiently they perform this function, the more power will be extracted from the fuel delivered.

Here's a pointed example. Raw fuel in an engine's exhaust gas is called "unburned hydrocarbons" (HC). From personal experience, we've seen controlled engine dyno tests where the only changes being made were to carburetor atomization efficiency . . . nothing else. By the singular adjustment to booster venturi design, power level changes on the order of 4 to 6 percent were recorded with the same amount of delivered air and fuel.

Surely, sizing of carburetors to specific engines and applications is important, as are all the precautions about making one work properly. But whatever you can do to improve mixture quality by atomization efficiency, anywhere from the time fuel leaves the carburetor to when it reaches the combustion space, is a path to increased power.

DENNIS WELLS' CARB DOS AND DON'TS

No stranger to the pages of *Circle Track*, Dennis Wells (Wells Racing Engines in Duncanville, Texas) is a wealth of practical information. His two-plus decades building racing engines of all types easily qualifies him as a valuable source for the Saturday-Night racer. Following is a condensed list of circle-track carburetor do's and don'ts.

Air Cleaners—"Clean and apply grease to the bottom *and* top of air filters to seal out dirt. Air bleeds can easily become plugged, leading to improper air/fuel mixtures and a lot of wasted time hunting elsewhere for the problem."

Carburetor Floats—"Periodically, check float height, leaks and deformation. Leaky floats upset fuel-bowl levels and impair main fuel delivery. Deformed or partially collapsed floats can also cause fuel level and calibration problems."

Needles and Seat—"Track environments are so dirt-hostile that regular checks to make sure 'trash' hasn't become lodged between seats and needles are important. Also, on a 2-bbl alcohol engine, remember that you have to force twice the amount of fuel through the needle and seat assembly as you do for gasoline. This is a critical area for providing adequate fuel supply."

Fuel Filtering/Delivery—"Install a 20-micron filter near the carburetor or fuel inlet. And for alcohol-fueled engines, don't ever use a paper filter. Use stainless only. Also, use a large filter on the fuel tank outlet. Install a fuel system shut-off valve within the driver's reach and always shut off the fuel before turning off the engine. And vacuum-check the 'suction' side of the fuel line."

Fuel Pressure—"Use no more than 7 pounds of fuel pressure idle. For gasoline, use 6 psi at maximum engine rpm. And if you're running alcohol, 9 to 11 psi at maximum rpm is the limit (depending upon jet size). Also, I've found that jet size and fuel pressure may vary, depending on track length, banking and who built the carburetor. Rather than experiment, I suggest going back to the builder if anything doesn't make sense regarding how the carburetor is performing."

Air Cleaner Design—"I've found that the base of an air cleaner assembly is critical. How the base directs air into the carburetor can have a material effect on engine performance, particularly 2-bbl applications where the carburetor tends to saturate (become a restriction) before maximum power is reached. Of course, the shape of the cleaner top and height of the filter element are important, but I recommend giving serious thought to the base's shape. For example, on the old-style Holley 500 2-bbls with a choke horn, we've found the K&N 'flow-thru' design worth 20 horsepower, a gain I think is largely due to the assembly's base design."

Carburetor Spacers—"As a rule, I think you need one and usually of open design. When one with holes or special shapes works, it often suggests a problem somewhere else in the engine isn't being addressed (too much camshaft, too large a carburetor, etc.). Generally, the smaller the carburetor, the higher it likes to be from the intake manifold's plenum floor. This is especially true if a two-plane manifold must be used. The floor of the upper plane can become a source of air/fuel separation at high rpm."

Fuel Bowl Vents—"Pay attention here. If bowl vents don't see atmospheric pressure all the time (or close to it), you can expect the overall carburetor calibration to mislead you, particularly when you're trying to read the effects of main jet changes. For example, if the bowl vents see pressure lower than atmospheric (often at high vehicle speeds), main system fuel delivery can go lean, giving the impression improper jets are installed. The opposite is true if the vents become 'pressurized' somewhere on the track. Work toward getting 'undisturbed' air around bowl vents. This is another reason why 'dyno jetting' and 'on-track' jetting don't always correlate."

Throttle Linkage Operation—"Be sure the throttle linkage stop is on the floorboard and doesn't use the carburetor as a stop. If you allow the carburetor linkage to be the stop, it'll prematurely wear out the throttle shafts. I like to make the throttle linkage open very slowly for

continued on next page

slick tracks. And if you really want to 'tune' the linkage, modify the linkage of the primary and secondary to open according to driving style. It shouldn't take much imagination to know what I'm talking about here."

Changing High-Speed Air Bleeds—"I know you can buy carburetors that have removable high-speed bleeds. My experience has been you should leave these alone unless you're tuning the engine on an inertia (wheel) dyno and are using both a fuel flow meter and a reliable O_2 sensor. These bleeds can really be sensitive, to the point that a 0.001-inch change can make a significant difference."

On Power Valves—"Check these parts *often*. Even the slightest backfire can damage a power valve. Generally, on 2-bbl applications, we like to stay in the 8.5–9.5 range. For 4-bbl applications, 6.5 is typical. These ranges apply to engines running either gasoline or alcohol. Before the new HP-style Holleys, power valves for their carburetors didn't have backfire provisions. Now they do."

Jetting for Alcohol—"We like to keep this pretty simple. What we do is start running the engine rich until it misfires in the upper rpm range. Then we lean the jets until the misfire goes away. You can, of course, tune more accurately than this, but it requires using equipment many racers don't have access to."

Chapter 16

CARB CONCEPTS

You Know the Basics, Now Let's Dig A Little Deeper

Text and Illustrations by Jim McFarland

Read along as we discuss carburetors and the peripheral issues, including companion parts, that may affect a carburetor's function. Hopefully, it will send you right to your toolbox—or local track.

What follows is a brief recap of carburetor circuits, functions, and intentions. The purpose of the review is to lay the groundwork for a dialogue of how other engine components may affect the ability of a carburetor to provide the proper air/fuel mixtures, especially in conditions encountered on the racetrack.

THE NOTION OF A PRESSURE DIFFERENTIAL

If there exists two different pressure values, a pressure differential is established. Typically, the higher of these two values will cause movement of whatever substance is contained in the path. In the case of an engine's carburetor, atmospheric pressure (or some artificial source) creates a force that seeks to equalize the pressure inside the cylinders with the atmosphere. This differential causes air to move into the engine and is produced by several factors, including piston displacement, rpm, valve timing, cylinder pressure, and size of the carburetor. In fact, we will study these components to determine their impact upon carburetor function.

THE ELASTICITY OF AIRFLOW

Here's an important concept. To explain, let's use the analogy of a string of railroad cars. As the locomotive begins to pull the train of cars, there is a chain reaction of movement beginning with the cars connected nearest the locomotive. Each car remains still until the car directly in front of it moves forward enough to take out the slack in the linkage.

The chain reaction continues rearward until the entire train is in motion. If we consider the pressure drop in an engine's cylinders to be analogous to the locomotive, then the first molecules of air to move into an engine begin at the intake valve and continue (the chain reaction) back to where air enters the engine. The reaction is complete when the pressure along the path is equal to the atmosphere.

This concept is important because it relates to how a carburetor senses air movement. In practice, airflow into the cylinders pulls outside air into the carburetor. How much time is required to complete the reaction is influenced by the physical distance the carburetor is from the cylinder. Notice we said *influenced.* Intake passage, cross-section area, valve timing, piston displacement, mechanical compression ratio, ignition spark timing (believe it or not), carburetor size, and rpm all have an effect on how quickly a carburetor recognizes reductions in cylinder pressure.

As an example, think about what happens when you increase the distance between a carburetor and its intake-manifold mounting flange. All else equal, it delivers less fuel, right? And this brings us to the next topic of discussion: carburetor signal.

CARBURETOR FUEL-METERING SIGNAL

There are various ways to define this condition. Once again, consider the notion of pressure differential. If two differing pressure conditions create flow, then we may say the greater the pressure difference, the greater the tendency to create flow. In "carburetor speak," the tendency to create flow can be called "signal." The higher the signal, the greater the pressure differential.

Since carburetors are essentially pressure-differential

This Holley Dorton "Signature Series" 390-cfm model has features that include screw-in air bleeds and emulsion jets, screw-in power-valve channel restrictions, and a four-corner idle system. Plus, it's NASCAR approved in certain classes of racing.

devices, their ability to move fuel into a passing airstream is a function of pressure differential or fuel-metering signal strength (plus the size of the orifice through which the fuel passes).

Inside the carburetor, a venturi increases carburetor-throat flow velocity while reducing the static pressure to which incoming fuel is exposed. The venturi utilizes an increased flow rate to aid fuel atomization. From the standpoint of an engine's combustion efficiency (the conversion of fuel into heat or power), breaking the fuel down into the smallest droplets possible—while keeping the droplet size consistent—is critical.

Since both air and fuel are subjected to the same degree of fuel-metering signal, a key characteristic of a good, power-producing carburetor is that it atomizes the fuel efficiently. If it was possible for a single carburetor circuit to maintain the ratio of air and fuel in proportion to changes in airflow, then only one circuit would be required. Unfortunately, it turns out that the rate of increase for airflow (as pressure increases) is greater than that for fuel. Because of this, carburetors must be designed to contain a network of circuits, each designed for the different pressure differentials associated with engine operation.

BASIC CARBURETOR CIRCUITS

We'll be brief because this is only a supporting segment of the story. However, an overview is necessary to understand the underlying effects other engine components can have on a carburetor's performance.

Idle Circuit

Under conditions of light or no load, such as idle, an engine receives mixtures from this circuit. As engine speed, load, or both are increased, the idle circuit provides additional fuel to support combustion. This circuit continues, providing additional fuel until it reaches a point of saturation, or maxes out. As rpm increases, the idle circuit continues at max output and is assisted in providing the engine fuel by other circuits.

Idle Transfer Fuel Source

Carburetors also have another source of fuel at idle. Sometimes called the idle transfer fuel source, these are small slots placed parallel to the carburetor's throats. As the throttle opens to an intermediate position, basic idle circuit sensitivity is reduced. Idle transfer fuel is provided through the slots to reduce or eliminate dead or flat spots during light throttle acceleration.

Intermediate Circuit

When an engine's load and rpm increase beyond idle conditions, additional fuel is necessary to maintain the correct mixture ratio, and the intermediate circuit is activated. Adding to the fuel still being provided by the idle circuit, intermediate fuel sustains combustion into a higher, or more loaded, set of conditions.

Main Fuel Supply Circuit

For carburetors not fitted with an intermediate circuit, a main supply activates to meet the needs of increased engine speed and load.

Accelerator Pump Circuit

During conditions when an engine's throttle is suddenly opened, it is necessary to provide a quick burst of fuel in order to maintain proper air/fuel ratios. This circuit typically consists of a fuel reserve (reservoir) that is displaced directly into the carburetor's throats by a pump or plunger.

Power Enrichment Circuit

Called by various names, this circuit provides fuel during high-load conditions. Carburetors use manifold vacuum (using a vacuum diaphragm) and some type of needle/seat arrangement to sense this condition. A decrease in intake manifold vacuum indicates an increase in engine load. At a preset vacuum level, the diaphragm valve opens, admitting fuel to maintain proper mixture ratios for the increased loading. Power-enrichment valves can be designed to function at whatever manifold vacuum levels suggest additional fuel is needed. Such valves can be designed in stages (typically two) for further refinement of fuel versus load conditions.

Air Bleeds

Fine-tuning carburetor circuits is possible by using air bleeds (calibrated orifices exposed to atmospheric pressure) and emulsifying (mixing with air) fuel throughout a range of engine speeds and loads.

Each carburetor manufacturer determines its methods and calibrations. However, contemporary racing carburetors often allow interchangeable air bleed orifices or emulsion tubes, adding further refinement to a racer's ability to optimize mixture ratios and power.

FACTORS THAT CAN AFFECT THE CARBURETOR SIGNAL

We have established that a carburetor is fundamentally a pressure-differential device, and its efficiency in providing fuel is affected by various changes in airflow. The following are mini-discussions about principle parts or equipment that can affect fuel-metering signals, air/fuel ratios, and ultimately, power.

Piston Displacement and Power Level

Let's assume we have a carburetor that has been calibrated for a 355ci engine producing 500 hp. If we place the same carburetor on an engine of 406 cid (even with the same horsepower), the air displaced per cylinder will increase proportionally and affect the carburetor's fuel metering signals. OK, that's a no-brainer; let's consider changes in volumetric efficiency.

Volumetric Efficiency

Here we have the same carburetor calibrated for an engine of 355 cid and 500 hp, and we make a change that increases volumetric efficiency (VE). Again, both air rate and volume of flow (per rpm) increase, which alters the fuel-metering signal. A change in VE may be the result of a number of factors, but generally a gain in net airflow (within some range of rpm) will increase VE. Keep in mind that you can also lose VE by decreasing airflow. A common example of this is oversized ports that penalize VE at low rpm but gain at higher rpm when flow rates increase. In these cases, calibrating your carburetor requires a little more creativity, but matching fuel flow to airflow (signal) is still the objective.

Jet Size and Flow Rate

A jet is a calibrated hole or orifice. Its rating is based on the amount of fuel it will flow as lower

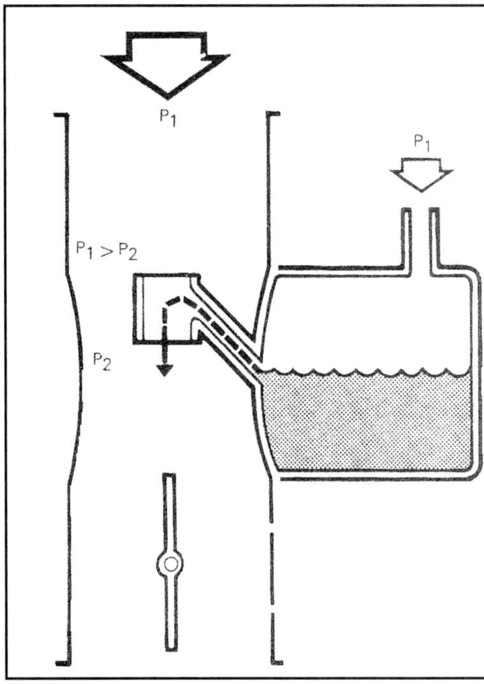

This illustration shows the importance of maintaining control of carburetor fuel bowl vent tube pressure. Fundamental pressure drop develops between P_1 (atmospheric pressure) acting on both the carburetor throat and fuel bowl. This pressure differential is a function of the relationship $P_1 > P_2$. As P_2 decreases, the differential between atmospheric and throat pressures increases, thereby increasing fuel flow at the carburetor's throat. Uncontrolled variations in bowl vent pressure disrupt proper carburetor function.

pressure is developed across the jet. As carburetor signals change, so does the amount of fuel that passes through the opening. The key here is the fact that jets flow more fuel in proportion to the pressure drop, until the jet reaches its saturation level.

What does this mean? Take a No. 80 Holley jet, for example. It will flow at less than capacity if the pressure drop it is experiencing is below its maximum flow rate. When this maximum is exceeded, a larger jet size is required. Keep this in mind when calibrating a carburetor for an engine's specific metering signals. If the signals are low, it may be necessary to install larger jets just to obtain proper air/fuel ratios.

Jet Size and Mechanical Compression Ratio

The reason to increase an engine's compression ratio is to increase the net cylinder pressure. Unless other changes are made (valve timing, exhaust port flow enhancement, and so on), residual pressure at the time of intake valve opening and exhaust blowdown will also increase. Either at the instant of intake opening (reversion spike) or during valve overlap, a potential change in pressure drop across the carburetor occurs. Again, a jet-size (or companion calibration) change may be required.

Demon Carburetion's Race Demon concept is available in a range of flow sizes and two versions: Race Demon and RS Race Demons. Race Demon features include removable boosters, CNC-machined billet-aluminum metering blocks, streamlined air entries, and sight-glass fuel bowls. RS Race Demons provide these features in addition to removable venturi sleeves.

In truth, the carburetor doesn't know the size of the engine it is mounted on. It only recognizes pressure drop and flow. It's a mixing valve, nothing more, and it needs to be calibrated with that in mind. As stated previously, the focus should be on combustion efficiency—specifically, obtaining the highest level of efficiency under the widest range of conditions.

Careful spark plug readings or observing an engine-dyno fuel curve will be your best indications.

Jet Size and Exhaust Backpressure

Be careful here. Changes to an engine's exhaust system that either increase or decrease net flow can be read by a carburetor. More specifically, let's assume we've reduced backpressure by whatever means available. This opens the option to reduce the amount of valve overlap that was required for efficient exhaust flow before the backpressure reduction. As far as the carburetor is concerned, the net effect of both changes may be that recalibration is not required. Chances are, the reduced backpressure will have improved combustion efficiency, in combination with less valve overlap that helps boost cylinder pressure. Under these conditions, a higher percentage of the fuel is being converted into power.

Jet Size vs. Carburetor Size vs. Piston Displacement

You have an 830-cfm carburetor that works well on a 355ci engine. Then you decide to run it on one that's 400 inches. All else being equal, and assuming the carburetor isn't limited by flow capacity, the 400-inch engine will ingest more air than the 355 at any given rpm. That's a given. The issue is how to deliver the correct air/fuel ratio for each engine based on the amount of air supplied.

The Importance of Bowl Vents

Don't overlook the importance of providing proper fuel bowl pressure under all circumstances of engine operation. Why is this important? Because the activation and operation of a carburetor's fuel delivery circuits include atmospheric pressure acting on the fuel bowl.

Remember, since a carburetor is designed to sense differential pressures, it relies upon relatively constant pressure at the bowl vent equal to the atmosphere. Now, there are times when this is not possible (underhood airflow conditions and so on), and jetting is selected to compensate for such deviations from atmospheric pressure. What is critical is that these abnormal pressures are relatively constant and do not fluctuate. The carburetor's circuits rely upon a stable bowl pressure. Placing carburetor bowl vents in areas where the pressure level changes can lead to lost power and, in some cases, damaged engine parts.

Carburetors and Spacers

The primary reason for using carburetor spacers is to improve the dynamics of wet flow (air and fuel) as it changes direction from the carburetor's exit to the manifold runners.

The goal is to improve the quality of the air/fuel mixture moving from the carburetor to the intake passages, and elevating the carburetor from the plenum chamber floor helps. You don't want to produce the effect best described as a hose squirting into a bucket.

But there is a secondary effect to moving a carburetor farther from the manifold runner entries. The fuel-metering signal tends to weaken, requiring larger jets with less resistance to fuel delivery. This is a restorative step intended to compensate for: 1) reduced fuel signal and 2) improved air/fuel mixture quality.

Don't forget to reconsider spark timing after the spacer is installed. If you've improved mixture quality, chances are good that the flame speed will have changed (typically faster), requiring an attending adjustment to spark timing.

SOME AFTERTHOUGHTS

Dirt is a fundamental enemy of carburetors. Good air and fuel filtration are essential to proper carburetor operation.

Float settings should be checked and/or adjusted to meet different track conditions. What works in one location may or may not work in another.

Keep an eye on all air-bleed openings to make certain they are clear. It doesn't take much residue to clog one of these bleeds, and the results can be less than pleasing—especially at high rpm.

Particularly for longer races, try to prevent heat from the intake manifold from saturating a carburetor. Aerated (vaporized) fuel can wreak havoc on fuel delivery.

A carburetor's ability to make power is not solely dependent upon flow capacity. Booster design and placement is critical. Both of these components are tied to efficient atomization of fuel and should be considered when choosing a particular brand or model. There are specific instances where a change in booster configuration or location with no alteration to air or fuel flow has shown power increases.

It's back to the subject of air/fuel mixture quality. We cannot overemphasize the importance of this issue, especially where carburetors are concerned. Providing sufficient quantities of air and fuel, properly timed, is one concern. Making certain that fuel is delivered in its most atomized state should be foremost in the entire induction system process.

Illustration Showing Relationship Between Venturi Booster & Carburetor Throat

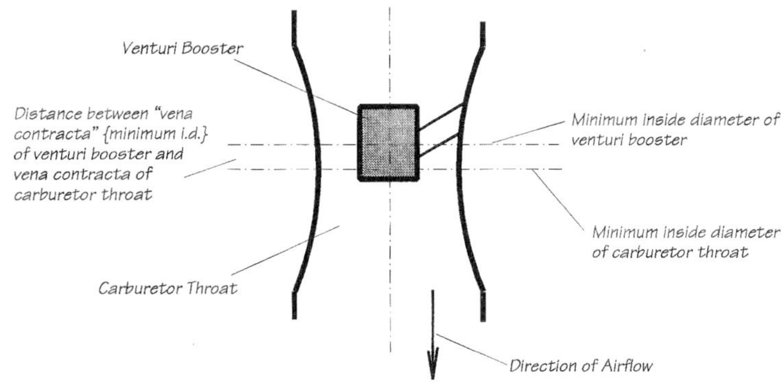

NOTE: Technically speaking, a "vena contracta" applies to the convergence of a free liquid passing through an orifice. In the case of carburetors, it may be applied to the minimum diameter of a throat {carburetor or venturi booster}, which produces the highest rate of flow, leading to the lowest static pressure region. In fact, this may or may not be at the physical minimum i.d. of a passage. It could be slightly beyond this point. In carburetors on normally-aspirated engines, this area produces the greatest pressure differential between throat/booster and atmospheric pressure and, therefore, the strongest fuel metering signal and opportunity for fuel atomization {in the case of venturi boosters.}

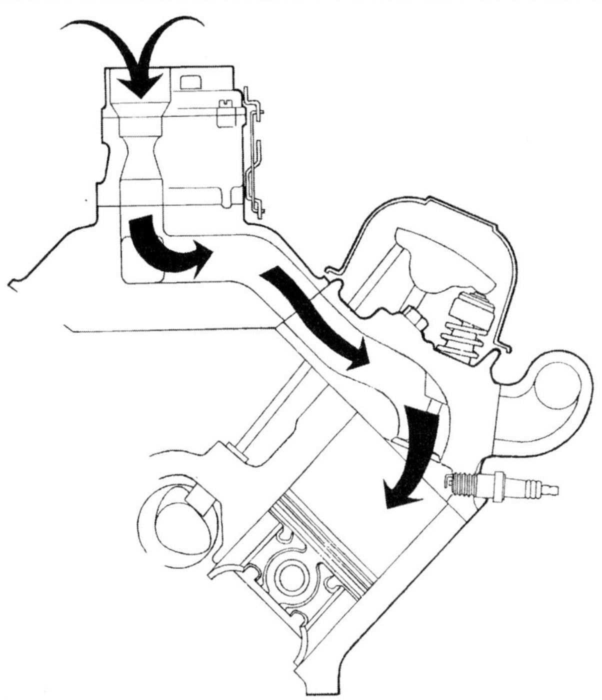

Illustration Showing "Elasticity of Flow" Concept

NOTE: Initial movement of inlet air begins at inlet valve and progresses back toward and to carburetor air inlet point. A brief time lag occurs between air movement at the intake valve and, subsequently, at the carburetor throats. Notwithstanding passage sizes and flow rates, the strength of this "metering signal" is a function of the distance between combustion space and carburetor throats. For this reason, jet size {carburetor calibration} requires adjustment when the distance is altered, by whatever means.

Chapter 17

HEADER TECH

These Parts Play a Major Role In Power and Efficiency

Text and Photography by Jim McFarland

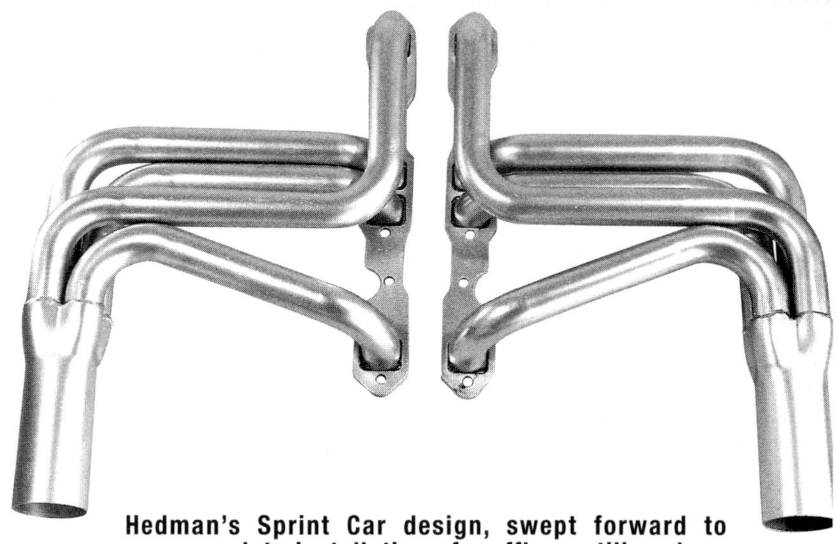

Hedman's Sprint Car design, swept forward to accommodate installation of mufflers, utilizes long primary pipes that develop and maintain mid- and upper-rpm power for engines that run in the 5,500–9,000 rpm range.

By design, this header chapter deals with exhaust gas removal from an engine's combustion space. It includes related topics such as valve timing and blowdown pressure, as each relates to the efficient removal of exhaust gas. You'll also get a taste for how basic header dimensions affect torque output, methods to use for matching headers to piston displacement and rpm, and the benefits of stepping primary pipes and changing collector volumes. While none of this information will place you in the category of expert, it'll certainly help you past the point of the "Gimme that set; I like its looks" approach.

WHAT HAPPENS BEFORE THE EXHAUST VALVE OPENS

Part of an engine's ability to produce power is a function of its volumetric efficiency (VE). We will define VE as a comparison of two quantities of air (or mass): the amount that passes into a cylinder during a single intake stroke compared to the amount that would occupy the same cylinder under conditions of atmospheric pressure (or inlet density). Typically, this comparison is expressed as a percentage. The higher the VE, the greater an engine's potential to produce power because there is more air for combustion.

All else being equal, increased VE produces higher air/fuel charge mixture density, resulting in faster flame travel (combustion process) and greater expansion pressure exerted on the piston. In fact, as combustion gas expands and produces work on the piston, its temperature begins to decrease. Frequently, this period of piston movement is called the "power stroke" but is also known as the "expansion stroke."

Regardless of how much energy is released as work on a piston, pressure in a cylinder (just prior to exhaust valve opening) is much higher than atmospheric. Keep in mind that even after the exhaust valve opens and flow begins outward to the atmosphere, cylinder pressure is typically higher than the intake manifold pressure at the time of intake valve opening. The more efficiently an exhaust system can lower cylinder pressure prior to IVO (intake valve opening), the less combustion residue will pass back into the intake system (so-called reversion). If the exhaust valve opens too soon, pressure in the form of work on a piston may be lost, and reversion energy will increase. If the exhaust valve opens too late, blowdown energy will be lost, and power will suffer from the dilution of fresh air/fuel charges.

THE BLOWDOWN CYCLE AND ITS EFFECTS ON POWER

If you can visualize atmospheric pressure working on the intake side of a normally aspirated engine, compare this analogy to the pressure contained in a cylinder (at the end of the expansion stroke) just prior to the exhaust valve opening. At this point, the exhaust valve opens (EVO) and cylinder pressure begins to decay into the exhaust system, including through the headers.

For the purposes of this chapter, let's define blowdown pressure as a pressure pattern (or history) developed from EVO to the time when pressure in the exhaust system equals cylinder pressure. Obviously, exhaust gas passing from the cylinder during blowdown does so through the release of stored energy. At the end of the blowdown period, any remaining exhaust gases must be displaced from the cylinder

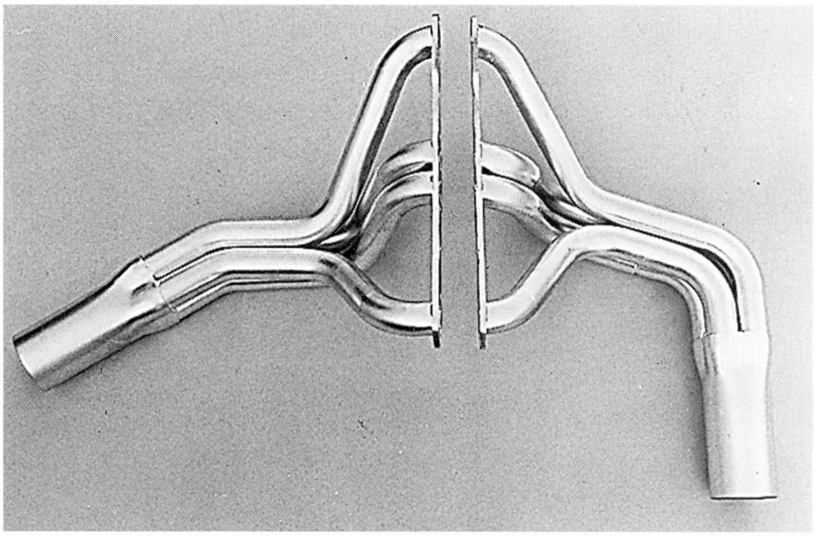

For USMS and IMCA Modifieds, these NASCAR-legal Hedman headers are designed for a specific chassis fit while intended to provide good off-corner torque, due largely to primary pipe length and not collector volume.

by piston motion. Of course, any residual exhaust gas contaminates the next fresh air/fuel charge.

blowdown efficiency is important to the integrity of fresh air/fuel mixtures, primarily because the inability to effectively remove combustion residue translates into potential power losses. Factors that affect blowdown efficiency include mechanical compression ratio, ignition-spark timing, heat losses to combustion-flame-exposed surfaces, combustion efficiency, EVO, and the rate of combustion (burn).

In particular, you will find that engines exhibiting rapid rates of flame travel have lower blowdown pressures than those with slower rates of combustion. One reason these are important factors is that valve timing should be matched to flame-travel characteristics. For example, as flame speed increases, valve-timing events can be shortened (especially EVO). Correspondingly, earlier IVO can occur since blowdown efficiency with quicker burns reduces residual cylinder pressure and tempers reversion conditions.

Why is all this important to the exhaust header function? Because even if combustion-pressure conditions and valve timing support good exhaust-gas delivery, a poorly selected (or designed) header can negate some or all the benefits that precede its receipt of combustion by-products. If you remember nothing else from this part of the chapter, know that blowdown pressure is instrumental in initiating the exhaust process and can affect combustion efficiency by inadequate removal of spent air/fuel charges.

BASICS OF THE EXHAUST GAS PROCESS

In general terms, the flow of exhaust gas into a header is not characterized by a slug of high pressure or high density of combustion residue. Rather, a set of waves is created whose excursions back and forth in the various exhaust passages comprise a complex system of dynamics. Even though there is a net flow of gas (or working fluid) through an engine's exhaust system, certain types of well-defined waves traverse these passages. The tuning of waves by specific exhaust passage dimensions, which include primary-pipe diameters and lengths, collector sizing, mufflers (if required), and tailpipes, are the subject (and object) of performance or racing exhaust systems. Headers, which are primary to such tuning efforts, are the focus of much design experimentation, study, and controversy.

Functionally, there are two types of exhaust pressure waves. Both of these are classified as finite amplitude waves. For purposes of clarification, finite amplitude waves are of significantly higher decibel (dB) than acoustic waves and, according to most reliable sources, are not practical for use in the tuning of either exhaust or intake systems.

In a typical exhaust path, there are two types of finite amplitude waves: a compression wave and an expansion wave. Compression waves initiate the flow in a header's primary pipe, traveling toward its entry into a collector or the next change in pipe diameter (volume). The former describes a primary pipe of constant diameter, the second a pipe of stepped or changed diameter.

Once a compression wave reaches the end of a primary pipe (or pipe step), an expansion wave is created that moves back toward the source of the compression wave (see illustration page 109). The movement (rate and timing) of these pulses or waves is important to efficient evacuation of an engine's cylinders; thus, it is critical to get the proper dimensions for pipes and collectors.

To visualize how these waves move through a field of exhaust gas (working fluid), consider corks floating on water through which waves are moving. Wave action will not carry the corks at the same rate of travel as the waves, yet there is motion imparted to the corks, which puts them in the same direction as the waves. The corks have velocity, but it is less than that of the waves. However, wave velocity imparts velocity to the corks, thereby establishing a relationship between exhaust gas or particle movement and wave movement within the gas.

SOME CALCULATIONS

At an engine's peak torque rpm, there is a flow rate (mean flow velocity) that can be measured in a header's primary pipes. Depending upon which method of calculation you accept, this rate is 240–260 ft-sec. In a header collector, it turns out to be about half these values (120–130 ft-sec). Assuming these values are reasonable, an equation can be developed that includes individual cylinder displacement, primary pipe cross-section area, and the rpm at which peak volumetric efficiency (or torque) is produced. What we're seeking to determine is the rpm at which we can expect a header's primary pipe to reach a peak value.

By applying some basic algebraic manipulations, the following can be solved for any of the different elements:

Peak torque rpm = (primary pipe area x 88,200) ÷ cylinder volume.

This approach enables you to determine the rpm at which peak torque can be expected (from the header system), given an engine of known size and headers you've already obtained. With some rearrangement of terms, this equation can become as follows:

Primary pipe area = (peak torque rpm x cylinder volume) ÷ 88,200.

In this case, you can determine the cross-section area of the primary pipes required to produce a torque peak at an rpm you want. Depending upon rear-gear selection, you may decide to obtain a header set that's dimensioned approximate to the rpm you'd like at torque peak and then fine-tune the combination with a rear gearset.

The point in all this arithmetic is that you can obtain (or work toward) a header sized to specific objectives. Given the fact that collectors function primarily below an engine's volumetric efficiency peak (as pointed out in the text), you can add collector volume to build torque below this rpm, and tailor out-of-the-corner power accordingly.

If you'd simply like to kill some bottom-end torque, either remove the collector or drastically shorten it.

Simplistically, speed up the waves and you'll speed up the gas particles. But it is the combined pressure excursions of compression and expansion waves that dictate net gas-particle movement in an exhaust passage. This brings us back to the importance of dimensioning header pipes, because their sizing bears influence upon efficient exhaust flow. In fact, with the proper dimensioning of an engine's exhaust pipes, exhaust-wave timing can coordinate with intake/exhaust valve events, rpm, and piston displacement to maximize volumetric efficiency. As previously stated, that's torque, and torque accelerates the vehicle . . . and we've come full circle.

The chassis-specific headers from Hooker are intended for NASCAR Late Model applications in which a broad power range is required.

CROSS TALK IN A HEADER SYSTEM

Further complicating the issue is what happens to pressure excursions when primary pipes are collected in a common volume (or collector). Now we no longer have a simple, "single degree of freedom" system (see illustration). As each primary pipe dumps into the common collector, additional pressure influences are registered by adjoining pipes. Also, depending upon an engine's firing order and primary pipe placement in the collector, multiple pressure activity is set up in the collector each time a cylinder fires.

For example, were you to locate a pressure transducer at the head end of one primary pipe in a system, pressure excursions would be recorded (during its inactivity) from wave motion in adjoining primary pipes activated by a firing cylinder. The point here is that the complexity of analyzing (and designing) contemporary exhaust headers is not a casual exercise. Even so, there are some greatly simplified methods that will approximate primary pipe size (cross-section area), independent of length, that include the influence of cylinder displacement, engine, and peak torque rpm. One mathematical approach to primary pipe sizing is provided in the accompanying sidebar. It's not dead-on, but it's surprisingly accurate for Saturday-night racer calculations.

RELATIONSHIP BETWEEN PRIMARY PIPE SIZE, RPM, CYLINDER DISPLACEMENT, AND TORQUE

Based upon the pressure excursion notion previously mentioned, flow rate in a header system is not constant, even at a steady rpm. So a mean flow velocity represents an approximate average flow velocity at peak torque rpm. As it turns out, the mean flow velocity (m.f.v.) in a header's primary pipe and the collector is generally accepted to be in the 240–260 ft-sec range. Given this information, it becomes desirable to configure a header (particularly primary pipes) that produces this m.f.v. at a predetermined rpm that fits a particular engine/track application. Stated another way, the ability to select pipe sizes that produce this desired m.f.v. at a specific rpm becomes a design tool in the construction of a header set.

Let's examine the aforementioned factors to see how each may affect the desired flow velocity. Obviously, changing pipe diameter will alter flow rate or velocity. At any rpm, enlargement of a pipe slows down the flow; area reduction speeds it up. Given a specific pipe diameter and rpm, increasing or decreasing cylinder volume will also change flow velocity in the pipe. Under these conditions, an increase in displacement will produce a corresponding increase in flow rate at the same rpm. Reduce cylinder size and flow rate will decrease. Since we know that flow rate (however it may be affected) is tied to volumetric efficiency, which in turn relates to torque, changing any of these variables (pipe diameter, rpm, or cylinder displacement) can affect the rpm at which peak torque is produced, seeking the desired m.f.v. at that point.

Here's an example. Let's say you have a set of four-into-one, V-8–type headers of 1.875 inches od (outside diameter). On the engine you're running, they produce a torque peak at 5,000 rpm. A larger header on the same engine will require a higher rpm at which to reach the same m.f.v., thereby shifting peak torque rpm upward. So if there's a method to all this madness, it seems to focus on determining the rpm (or rpm range) where the engine will be required to deliver its greatest torque output, and then (in the case of headers) sizing the parts to produce the m.f.v. associated with peak torque in that area of rpm.

The same approach can be applied to other areas of engine modification. For example, if a connecting rod length change is anticipated, knowing that this tends to change the rpm spread between peak torque and peak power, the header pipe size can be adjusted to support the change in rod length. But again, the central issue is critical rpm identification and getting the engine to perform where it's needed.

RECAPPING SOME THOUGHTS

First of all, spend some time studying the accompanying illustrations. They are intended to extend concepts advanced in this chapter and are necessary for a more complete understanding of what is sometimes obscure in text form. Next, realize that this information is not intended as a run-of-the-mill regurgitation of what may have been previously read or heard. It is intended to provoke thought and a perspective about headers that will enable a more accurate selection or modification process when it's time to choose a set for your particular application. You may need to read this material more than once. But if you come away from these pages with a clearer understanding of how headers relate to the overall engine package's function, then you've reached the goal.

Here are some quick-check guidelines in the selection and/or use of a typical collected header system:

Simplified Illustration of Compression & Expansion Waves

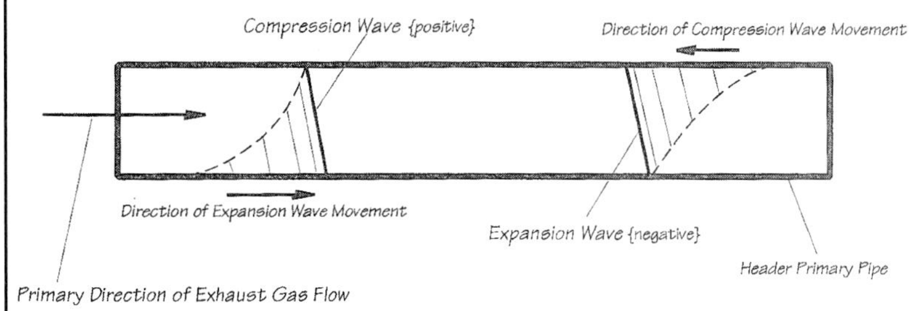

NOTE: As pointed out in the text, both compression and expansion waves traverse a passageway, decaying in a uniform fashion between cylinder firings. Exhaust gas flow occurs during and is aided by the action of both compression and expansion waves. Other than the influences caused by engine rpm, piston displacement, valve timing and related cylinder pressure-influencing varialbes, exhaust passage dimensions {cross section area and length} affect the dynamics of flow and "tuning."

Simplified Illustration of Single- and Two-degree of Freedom Primary Pipe Design

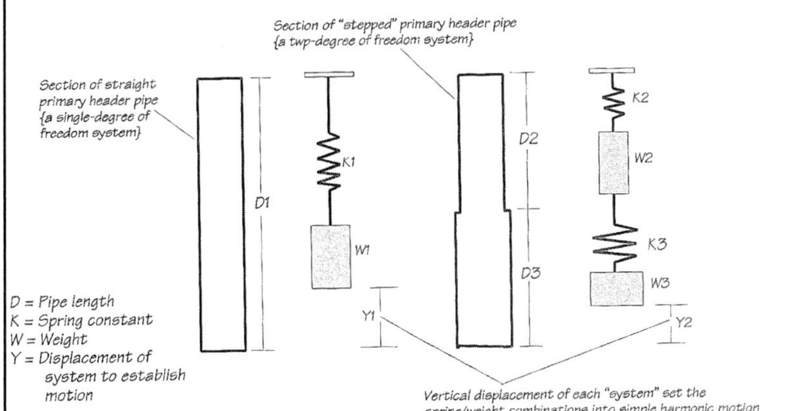

NOTE: These two examples represent "single-degree of freedom" systems and "two-degree of freedom" systems applicable to a conventional header exhaust. Although classifying these two pressure excursion systems as simple harmonic motion {depicted by the spring/weight examples} is not specifically accurate, the suggestion is that pressure movement in an exhaust pipe can be visualized as an oscillating sprang assembly. In a two-degree of freedom primary pipe system, two torque "boosts" can be accomplished, each developed at the point where the desired m.f.v. is created.

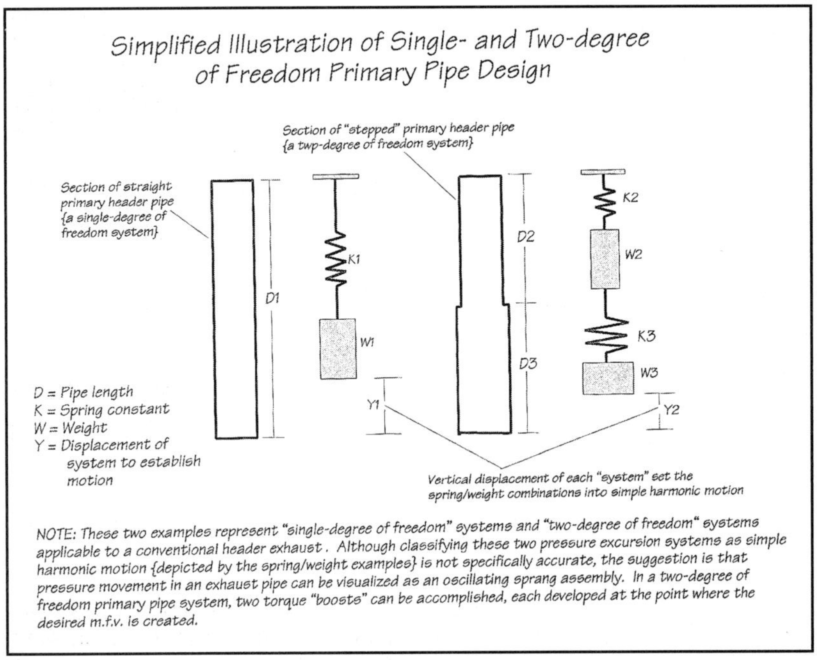

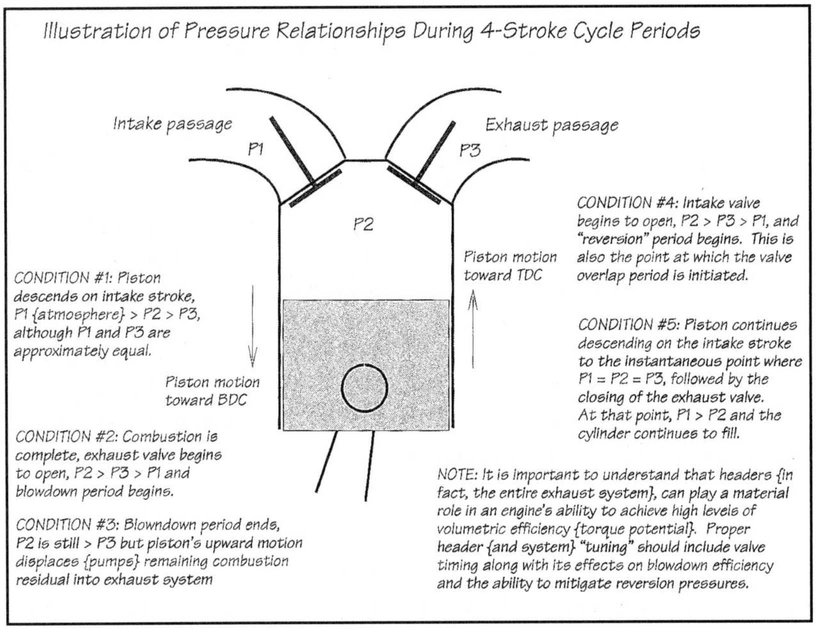

1. Increasing or decreasing primary pipe length tends to rock a torque curve about its peak rpm point. (It does not shift peak torque to a higher or lower rpm.)

2. The value of a header collector is essentially lost above peak torque rpm. Stated another way, collectors work at and below peak torque.

3. Peak torque rpm can be changed (insofar as an exhaust header's contribution to torque is concerned) by increasing or decreasing pipe size. Larger pipes shift peak torque higher; smaller pipes cause it to be lowered.

4. Crossover pipes in a dual-collector system tend to increase torque below peak torque rpm because this represents an increase in collector volume (simply more collector).

5. Building a header system with more-than-one-diameter primary pipe size causes a torque curve to be flattened over a wider range of rpm. (This is good if you want to broaden the effective range of a torque curve.)

6. Headers with stepped primary pipes tend to broaden the range of a given torque curve (see illustration).

7. Headers and intake manifolds (as systems) can be tuned, or produce separate torque peaks, at different rpm.

8. The influence of valve timing on headers is important. For example, as improved exhaust breathing is provided by a given camshaft, headers should be adapted to the rpm range where that improvement occurs.

9. As header efficiency improves (backpressure is reduced), combustion efficiency tends to improve. This provides an opportunity for more rapid flame travel, thereby reducing the amount of initial spark timing required. Work done on a piston approaching TDC on compression and during initial combustion (negative work) can be reduced by delaying initial spark as much as possible. Reduced combustion contamination from companion reductions in backpressure help speed up the burn and allow for less spark timing—all of which can achieve net gains in torque output.

10. Try not to over-size headers. In an attempt to make big peak-power numbers, it's possible to lose sight of the importance torque plays in the on-track performance of a circle-track car. Be realistic about the rpm range in which the engine spends most of its time—or needs to spend. Off-the-corner torque has won a truck-full of races.

11. None of the foregoing is cast in stone, but each can save hours of experimentation and expense if carefully considered.

Chapter 18

POWER MUFFLERS

Installing A Muffler On Your Race Car

Text and Illustrations
by Jim McFarland

With increasing frequency (no pun intended), racetracks are requiring their competitors to use mufflers on their cars. Here's a look into the physics of how this can be accomplished with minimum power reduction—in some cases, perhaps maybe even an increase.

To understand the complexities of an automotive muffler, particularly one intended for racing or high-performance engines, it's necessary to include some basic informational components. Expansion pressures tailing the end of combustion are emitted at energy levels much higher than atmospheric pressure, often resulting in objectionable sounds. These can be pleasant or not. For racing engines, it's becoming increasingly not, especially in areas with growing environmental concerns.

SOUND AND FLOW

Here's an analogy to consider: Let's say you are standing at the edge of a creek or any flowing stream of water. The moving liquid has weight (or mass) and velocity, therefore it displays kinetic energy (K.E.). How much K.E. it contains depends on the quantity of both these elements: mass and velocity. In the exhaust flow of a running engine, our example is now exhaust gas, and the amount of K.E. it displays is still a function of its mass and velocity. Simple enough.

Now, suppose you throw a rock into the water. Ripples are created, which are carried along with the movement of the water. These disturbances are like sound waves in the flow of exhaust gases.

Continuing our example, let's assume a piece of screen wire is placed across the stream just below the rock's splash. As the waves pass through the screen mesh, some energy is dissipated, and the waves that emerge on the other side are of a different size and shape—typically reduced. There is an energy exchange, or transformation, as the waves interact with the screen wire, resulting in a conversion of energy from one form into another.

That, basically, is what happens when high pressure, disturbed, heated exhaust gases leave an engine's combustion space and enter a "muffled" exhaust system.

Technically, sound may be defined as an alteration of pressure or particle displacement propagated in an elastic medium, producing variations in pressure. In a practical sense, an engine's exhaust sound can be either pleasing, of no consequence, or damaging. Its energy can be diminished or absorbed when passing through a medium or by hitting against a surface. Both can be and are used as automotive mufflers, racing or otherwise.

Sound in a Vacuum?

If you have an atmosphere inside a container (in this case, a muffler), then a medium exists through which sound waves can be propagated. As atmospheric pressure is decreased, the density of this medium is reduced, which diminishes its ability to move sound waves. While a pure vacuum is unlikely to be found in a muffler, the ability to establish and maintain low-pressure areas within the muffler effectively increases its ability to deaden sound. It is possible to design interior portions of a muffler that accomplish this objective.

Sound Frequency and Amplitude

Strictly speaking, we may define acoustical frequency as the number of sound waves (per second) passing any point in a sound field (any medium through which sound is moving). Units of sound frequency are typically measured in cycles per second. Since sound is an auditory sensation, the conditions that describe its presence relate specifically to what we can hear.

For example, the strength of a sound may be termed as its

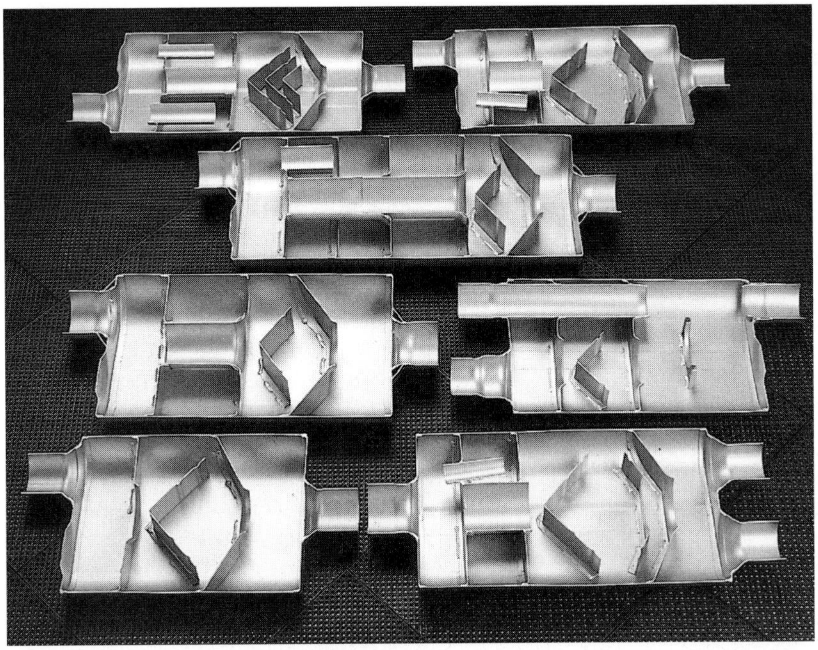

Here's an assortment of Flowmaster muffler designs that utilize sound frequency and amplitude modification. Note the use of Helmholtz resonators in some of the configurations (the upper left section utilizes two).

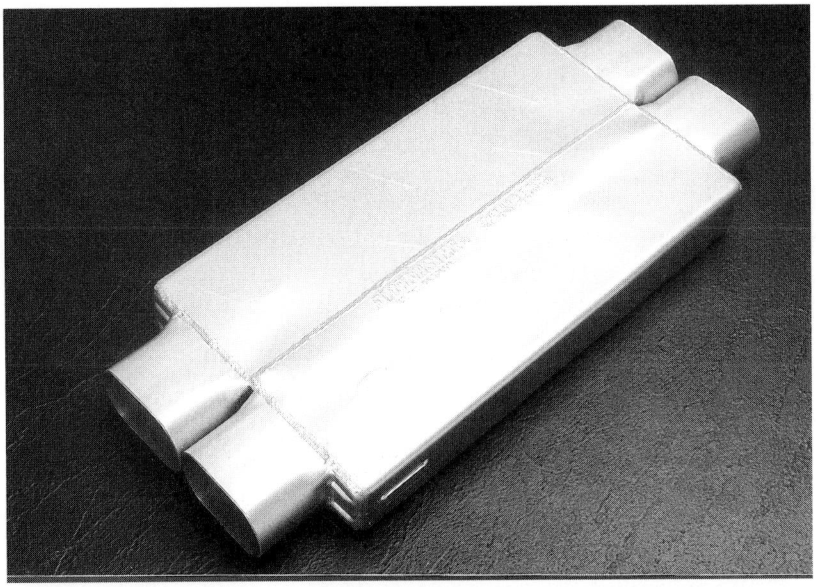

Developed exclusively for NASCAR Busch North cars, this limited-clearance Flowmaster unit is designed for applications requiring dual inlet/outlet muffler systems with oval tubing. Flowmaster also claims this design improves scavenging of the exhaust gases from the cylinders.

amplitude. High-strength sounds have high amplitudes. Combinations of high frequency and high amplitude are common in automotive mufflers, particularly on racing engines. Sound also travels faster in a heated medium than cooler ones. As a result, mufflers on racing engines are exposed to high-frequency, high-amplitude sound waves in hot exhaust gases, all of which increase the complexity of muffler design.

There is also the issue of frequency change that can be addressed in a muffler. Since certain sound frequencies are more pleasant (or less damaging to the ear) than others, it is possible to design mufflers that cancel (or alter) specific sound frequencies while allowing others to remain in the mix of exhaust sounds.

It may surprise many racers out there to learn that it's possible to design mufflers to reduce specific sound frequencies—while still providing low backpressure conditions.

Networks of baffles, instead of a combination of baffles and packing material, frequently characterize this type of muffler. For purposes of this discussion, baffles may include dimpled or perforated tubing, as well as indentations resembling louvers that either absorb or reflect pressure waves.

A device called a Helmholtz resonator is often used to change the tone (frequency and amplitude) of exhaust pulses. By definition, this type of resonator is characterized by an enclosure that communicates with outside pressure conditions by way of a small opening, usually a tube (see Helmholtz photo at left). Based on the volume of the resonator and the communicating tube, Helmholtz resonators will resonate at a single frequency, thereby affecting exhaust tone at that frequency.

EXHAUST BACKPRESSURE

Historically, the muffling or attenuation of engine exhaust sounds has been accomplished by adding a system that also hinders the free flow of exhaust gases, increasing backpressure. Low backpressure is important because we're trying to rid an engine of combustion byproducts—particles, or residuals, that don't tend to burn a second time.

An inefficient exhaust system (muffled or not) can cause these residuals to remain in the combustion space, not only preventing the entrance of fresh air/fuel mixtures on the next cycle but also reducing combustion heat via fuel-charge dilution. While there may be instances when a measure of backpressure is desirable, particularly during low rpm and to prevent "overscavenging" of cylinders, this is not the case in a majority of situations. Backpressure also causes natural EGR (exhaust gas recirculation). This reduces combustion temperatures and oxides of nitrogen, both of which are counterproductive to optimum power. In a racing engine, avoidance of combustion contamination is one of the rules to producing power. So we don't want racing mufflers that promote or cause backpressure.

Backpressure is also affected by piston displacement, power output, and rpm. All things being constant, backpressure (and exhaust sound) increases with rising rpm. It is important to keep all these elements in mind when selecting a muffler system.

OPEN PIPES VS. MUFFLERS

It is the net of pressure excursions (cycle) traversing an engine exhaust system that affect its output of sound and power. The notion that an open exhaust always makes the most power is not necessarily true. Examples abound, but here's one that gets to the core of the issue:

Consider a single-cylinder engine running at a constant rpm. Its exhaust system is a simple length of open pipe. Visually, there appear to be no restrictions to flow. However, by lengthening or shortening this pipe, we discover a corresponding change in rpm. If no physical impediment (e.g., a muffler) has been added to the exhaust's path, how can the engine's rpm change by merely altering the pipe length? Simple—the longer pipe through which the exhaust gases had to travel increased backpressure.

So, what if we did install an energy-altering device (possibly a muffler) that aided net flow while treating sound pressures as a separate issue? Now we're getting into the realm of exhaust tuning by lowering the effective backpressure created by an open exhaust system. By comparison, it can be said that a muffler with this design objective makes horsepower over an open exhaust.

In instances involving exhaust tuning of this type, a reduction in pressure at the system's exit amounts to reduced backpressure. If a muffler is designed to reduce

Schematic Illustration of a Helmholtz resonator

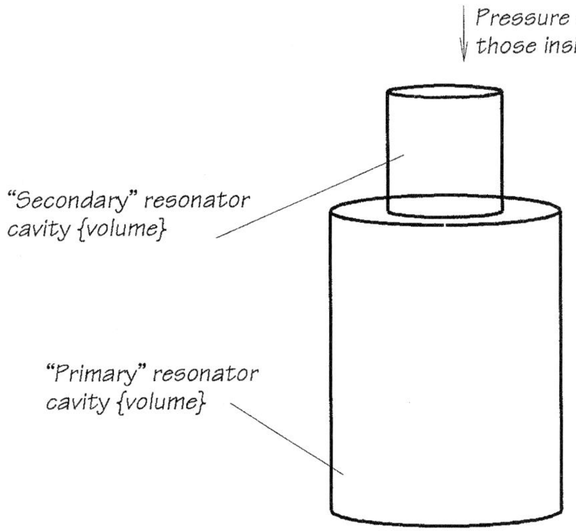

NOTE: By definition, a Helmholtz resonator is an enclosure of specific volume communicating with an external medium through an opening of smaller cross section area, resonating at a single frequency based on enclosure and port dimensions. In the case of a muffler, sound frequencies can be altered by communicating through the "secondary" cavity into the "primary" cavity. Tuned to address specific sound frequencies, based on dimensions, Helmholtz resonators are used to produce certain exhaust "tones" or "adjusted frequencies" consistent with required or desired sound characteristics. Pulsating exhaust pulses act against the captured "primary" volume, communicated through the "secondary" volume. While the shapes of these Helmholtz devices vary among muffler manufacturers, their function remains the same as described.

Schematic of Muffled Exhaust System Components and Importance of Component Placement Within System

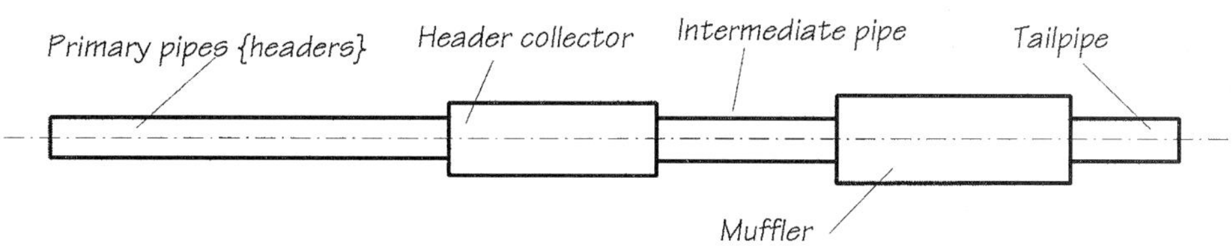

NOTE: As exhaust pressure pulses pass through this type system, downstream pressure conditions are altered each time a pulse experiences a change in path cross section area {or volume}. Upon so doing, a system of waves {expansion and pressure} begin traversing each successive volume {section} change...such as depicted here in the forms of header pipe, collector, muffler, tailpipe, etc. While the strength of these pulses includes influences from piston displacement, rpm and exhaust expansion pressure, {blow-down pressure} placement of the in-stream components can also affect overall exhaust "tuning." Therefore, such variables as muffler location, dimensions of tailpipe and distance of muffler from header collector can all influence net power. Comments about muffler placement are contained in the story.

Schematic Illustration of Typical Perforated tube/packed Muffler

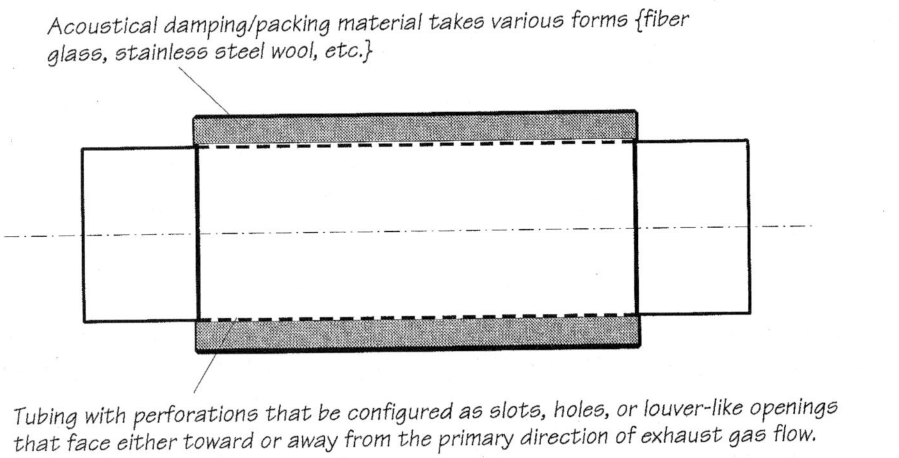

NOTE: So-called "straight through" mufflers may be designed with or without various types of packing materials, all intended to modify sound pressure and frequency during passage of pulsating exhaust flow. Numerous factors affect the efficiency of such mufflers, including total surface area over which exhaust gases pass.

its exit pressure below that of its inlet pressure, the device becomes a sort of one-way valve, which benefits combustion efficiency through reduced contamination in the combustion space. However, such a muffler must also prevent atmospheric pressure from causing flow backwards during times of low muffler pressure—like during valve overlap—if it is to qualify as a "one-way" valve.

MUFFLERS, BACKPRESSURE, AND VALVE OVERLAP

If backpressure is reduced by installing more efficient mufflers, then the opportunity exists to revisit the issue of valve overlap and how decreasing it may increase power. The same holds true in reverse if you're installing mufflers for the first time or even putting them on and off with regularity. Some valve-timing changes may be in order.

If the term "scavenge" is appropriate (scavenge is the effect of low-pressure conditions in the exhaust path pulling the burned contents of the cylinder into the exhaust system as the exhaust valve opens), its efficiency is said to improve with increases in overlap. During this period in valve events, atmospheric pressure can "see" all the way from where air enters to where it leaves the engine. This can wreak havoc on carburetor calibration and net volumetric efficiency (or torque). The addition of a muffler can not only affect carburetor fuel-metering signals, but its effects on combustion efficiency also cause changes in initial ignition spark timing. Typically, as backpressure is increased, fuel-metering signals weaken, combustion efficiency falls off, and more spark timing is required to ignite and sustain the burn. In addition, it may be necessary to increase valve overlap and/or retard the cam for power at higher rpm where contamination becomes worse (less time to evacuate cylinders of residue).

The corollary to these conditions is what happens when backpressure is reduced. In this instance, combustion efficiency stands to improve. But what can happen is cylinders may become "overscavenged," netting a reduction in volumetric efficiency because fresh air/fuel charges are passed through the combustion space during overlap. This possible loss of power can be addressed by shortening the overlap period, particularly closing the intake valve.

In addition, due to the potential for improved combustion efficiency (a reduction in air/fuel charge contamination), it's often possible to reduce ignition spark timing. Consider this: By delaying the point of ignition, less negative torque is produced at the crank, culminating in a higher net torque (increased IMEP) and more low-rpm performance—particularly for accelerating out of corners and when passing.

So it's not just about mufflers and their sound-attenuating properties. Companion parts and tuning considerations are key to making the most of a muffled situation.

If you give proper attention to the engine components that are affected by the installation of a muffler, it's possible to minimize power loss because of increasing backpressure. Overall, a race engine should be an integrated package of parts that work toward optimizing volumetric efficiency in the typical rpm range that is seen in a race. Properly developed torque ranges contribute more to on-track success far more than top-end power numbers do. There's no reason why the installation of a muffler needs to detract from that concept. It's just good, sound advice.

CRITERIA FOR SELECTING SATURDAY-NIGHT RACING MUFFLERS

Touchy subject here, largely because of the many variables involved. But here are some guidelines:

• If you're faced with muffling a race engine, find out what sound limits are imposed. There's no reason to reduce sound below what's necessary, particularly since you have some choices among mufflers to select.

• Let's suppose you're racing in an area (or class) that requires the use of a spec muffler. If this is the case, the competitive playing field gets leveled out. In this scenario, it's critical that you optimize all other aspects of the engine as outlined in this chapter.

• Once a muffler system is installed, keep an eye on engine coolant temperature. While backpressure can contribute to contaminated combustion and reduced flame temperature, it also tends to make the engine retain heat in the engine block and heads.

• Unless the muffler you've installed allows a particularly low backpressure, don't be surprised if on-track fuel economy declines—especially if you find additional jetting is required to restore performance.

• Mufflers that allow low backpressure virtually require you to rethink valve timing, particularly valve overlap periods. Shorter is the norm in many cases.

• Wherever possible, use a crossover pipe between collectors. This provides sound-pulse damping and a measure of noise reduction. You may find that a less restrictive muffler in conjunction with a crossover pipe will help you achieve required sound levels and maintain open-pipe performance.

• Overall, a muffler is an additional tuning volume placed in the exhaust system. If you can, experiment with its placement along the tailpipe. You likely will find a location that's superior to others along the length. One way to determine this location is to paint the muffler, run the car awhile, then find an area where the least amount of paint has been burned from the case. This is a signal for a location that minimizes the backpressure of the muffler you're using.

• In the vast majority of cases, a length of tailpipe extending past the end of a muffler will aid in its performance, both in sound attenuation and its effects on power. Typically, both are good.

Chapter 19

OIL SYSTEM TECH

Smokey Explains The Basics of This Critical Element

By Smokey Yunick

This engine is equipped with a dry-sump oil pump, the choice of many racers. Crankcase vapors are vented through a breather in a remote tank. The choice of oiling system runs hand-in-hand with application.

The lubrication of a race engine, especially the 50-year-old production passenger car engines that are used in stock car racing, has always been and continues to be tackled in an expensive and time-consuming cut-and-try method. Time and time again, racers apply shortsighted fixes, hot track tips, aftermarket parts, producer recommendations, and high-performance parts offered by the OEMs to keep their engines oiled.

For new racers, this learning curve is especially expensive and frustrating, which hurts racing in general. I'm going to help you new racers avoid the 50-year engine oiling learning curve you face by explaining what the common lubrication problems are and how you as a racer can handle them.

To start, we need to understand what oil itself is. And I don't mean so we can make it, but so we can decide what oil you need for your application. Being familiar with the different systems (wet sump, dry sump, accumulators, splash) available to lubricate your engine for your type of racing is also part of what we'll discuss. Often, the rules you race under force you to compete with a marginal lubrication system, and I can tell you what the limits of that system are. Filtration and cleanliness need to be achieved because they directly affect engine reliability. Oil is contaminated by many substances besides dirt, like water, air, and fuel, so understanding how these substances get in the oil and how to stop them is critical to durability.

Clearances and oil hole sizing, along with the overall routing of the oil through the engine, directly affect the ability of the oiling system, so they need to be addressed here. Optimum oil temperature and the venting of the oiling system are critical to extracting all the performance from the oil. Also, you need some basics to follow when troubleshooting an oiling problem, like how to examine engine remains and figure out a fix. And finally, being able to determine what products you don't need will be discussed. Some oil, additives, filtration systems, and other gadgets might not help you, but they could possibly hurt your engine, so being able to determine what will help your oiling system is something that needs to be discussed.

CHOOSING YOUR OIL

To pick the oil you use based on cost (whether you pay for it or get it for nothing) is a big mistake. One failed engine can be more expensive than paying for the oil for five years. In racing today, the biggest enemy is heat, so you need an oil that can maintain its properties at extreme temperatures. The three types of oil available to racers are bean oil, mineral oil, and synthetic oil. All of these oils deal with heat very differently. Bean oil, surprisingly, is probably the best lubricant for five minutes, but it can't hack it in a 15-minute race, so it's out 'cept for use in the draggers. Mineral oil is what you've been raised on. This is what most oils are, a straight petroleum-based oil with an additive package, either as a single viscosity or multi-viscosity (SAE 50 versus SAE 15W-50) oil. For years, racers feared the multi-viscosity lubes, but the truth is that the multi-viscosity has some advantages and no negatives I know of.

I've said this many times, to quote Kettering who was a famous GM engineer, "The price of progress is trouble." This was very true in early Indy and other forms of racing with regard to oiling. Bean oil or castor oil were the only lubes a racer could use. You could smell Indy from 5 miles out in May. It was an odor I really liked. Hell, we might have gotten high from it.

While the bean oil was the best for that time, it really wasn't good enough. Even so, it took the mineral oil companies years at Indy to win a race. They had to furnish engines to competitors to use their products. Hell, some racers stayed on bean oil 20 more years, long after it was a second-class lube! I mention that because we've been in the same type of transition since 1975. Man-made or synthetic oil, which is made of reformed hydrocarbons or even fats and acid, is now a much better lube than mineral oil. All the hot dogs in high-buck racing use it now. But many good racers are still resisting. Not all synthetic oils are good for racing, and they have given synthetics a bad name in racing. If I were to race anything today, I'd use Mobil 1 15W-50, and I want you to know Mobil has never given me so much as 1 quart of synthetic oil free or paid me a dime. Mobil has given me mineral oil free and one time a racing jacket, but that's it. I'm sure there are other synthetics that are just as good, but I haven't tested them. Basically, synthetics will continue to become the standard, so as a racer, learn as much as you can about them so you can make the switch.

Oil choice is one of the important considerations to the success of an oiling system. Specially made racing oils have proven more hearty than passenger car oil when considering the demands of a racing engine.

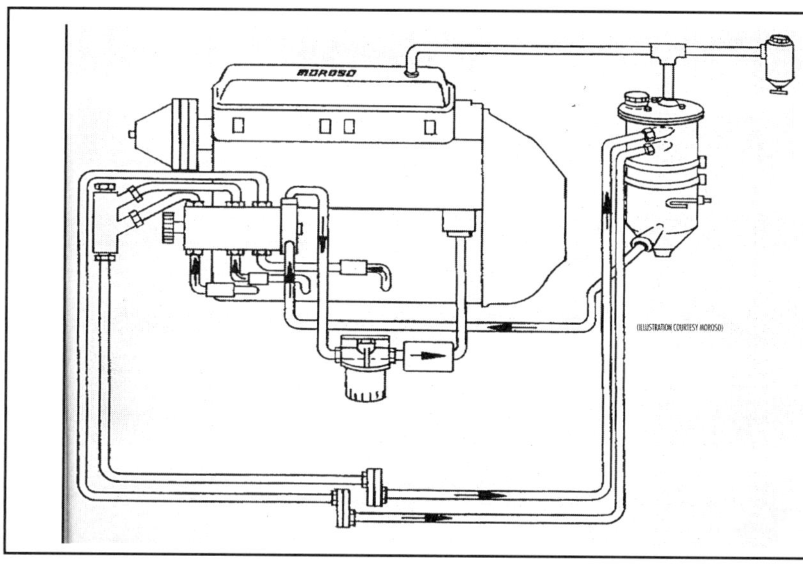

This is a simple explanation of the routing of a dry-sump system.

CHOOSING AN OILING SYSTEM

When you are deciding which system you are going to use to lubricate your engine, you should first check the track rules where you are going to race. There are two general classifications of oiling systems: wet sump and dry sump. I doubt anyone disagrees that a dry-sump system is by far the best, but you do what you have to do. The initial cost of the dry sump is higher, but in a season it would be cheaper because of better oil control and management.

If you are allowed a dry sump, you should use a pump with an air-separator section. Every high-buck engine running Indy or Formula 1 uses an oil-separation system.

Now if you have to run a wet sump, it can be done successfully. Your main enemy is oil starvation during cornering, acceleration, and braking. The second enemy is oil temperature. You can run an oil cooler and fight the heat, but you deal with less oil compared to a dry-sump system. A dry sump has the oil cooler and the sump itself, which also acts as a cooler. In running a wet sump, if the rules allow, you should run an oil accumulator. If you do uncover the pickup, the automatic intervention of the accumulator rescues you. Regardless of what oil system you use, don't try to cobble up oil pan sump tanks or oil pumps by yourself. Man, my learning curve on that took years and cost a fortune. As cheap as those parts are compared to the material and man-hours it takes to make 'em, you should at least buy your first one from a leading aftermarket vendor like Moroso. Then you can attempt duplication and/or improvements if you still feel the need.

There are many types of oil filters, and you often get what you pay for. This is not a place to cut corners for cost sake unless you want to replace an engine.

FILTRATION AND CLEANLINESS

Good filtration is a must, but this topic is loaded with bull and false claims. Assuming you have built your engine clean, all the plumbing and oiling system parts are clean, add clean oil, and have a good air cleaner (paper type) installed leak proof, how the hell much dirt is there in a race to hurt an engine? The only thing that can eat up bearings are pieces wearing or breaking off of inner engine guts. There is very little wear in a good engine, even in a 500-mile race. Most racers and filter manufacturers filter the oil too finely. OK, what does that hurt, you ask? Just this: the finer the filtration, the larger the pressure drop you get and also a reduction in flow rate. Well, you can make that up by increasing the oil pump pressure. What happens next is the added pressure heats the oil further and more oil is bypassed. This also uses (wastes) horsepower.

The filter construction and its area size and fitting size have to be considered. You can't get too big here. You damn sure can get too small. Too small means a big pressure drop. Try to do the job with one proper size filter, not two parallel filters. Never plumb filters in series—that causes a double pressure drop.

Look at the big filter. The best filter is one where you can remove a door and look at the collected residue, or something like that where you can examine the filter element (like a System 2 or Oberg). Take it apart and look at the filter after practice and before a race. It'll save you money and surprise race failures. I don't care how good an engine builder is—he has had some blown engines. The better an engine builder is, the lower his engine failure rate is. When you use a can-type filter, cut it apart and look at it. This is just a little more involved and occasionally spills a little blood. In my adventures, I'm sure various types of filters are better than others in the amount of air they put into oil. It's not uncommon for oil to be half air after 20 or 30 minutes of racing. So then you are running half air and half oil for lube. Air ain't worth a damn for cooling, bearing lubrication, or for pressure. I believe wire-mesh filters put less air into oil than paper filters. It's possible paper takes finer particles out of the oil, but the question is, do we need to filter that fine? Which is the lesser of the two evils? I use wire-mesh.

CHOOSING A PUMP

After you choose your oiling system (or are mandated by the rules with no choice), there are still many modifications you can do that can add up to success or failure in any and all of 'em. First, choose a pump that's the correct size for your application. For example, dry-sump systems are usually too damn big for the job. Here's how it hurts: The bigger the pump, the more power it takes to turn, the more air it puts in the oil pickup line, and the higher the oil temperature will be—same with a wet-sump oil pump. Those wet-sump

Oil pumps can come in different stages. This is a five-stage model, which will provide increased efficiency and reduce oil temperature.

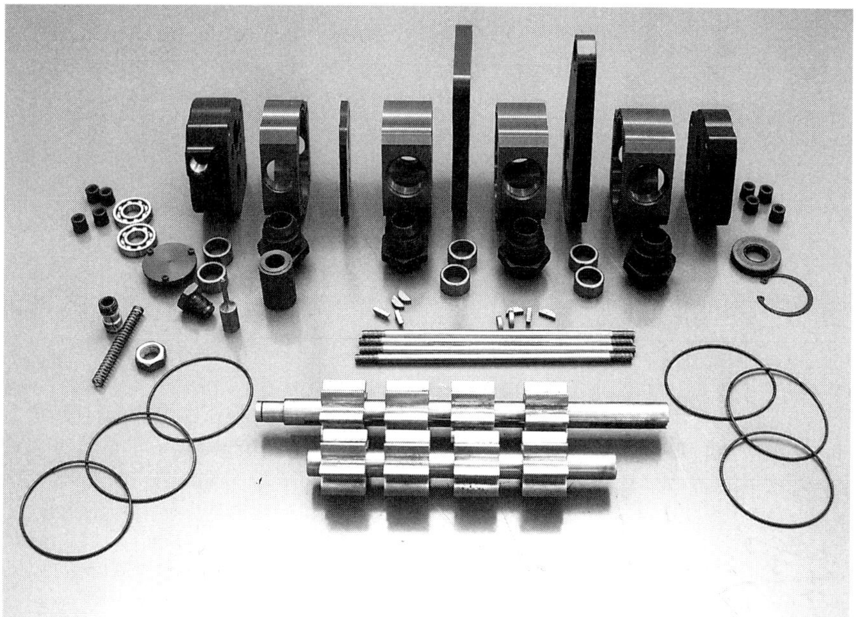

An exploded view of a dry-sump pump. It's a four-stage model from Fast Times Fabrication. Each of the small parts plays a critical role for the four stages of this pump.

big pumps bypass half the oil at high rpm and starve for oil because of the small pickup tube, which allows bearing damage. For example, if you take the trouble to put a 5⁄8-inch inside diameter pickup tube in a good small-block Chevy pump, it's better than a big-block pump.

It takes 8 or 9 gallons per minute (gpm) to run a small-block at 7,500 rpm, and the stock pump can do that easily. The shape of the oil pickup is very important, as is its distance from the pan bottom. You need at least a 5⁄8-inch diameter pickup tube. If the pickup is the wrong shape or too close to the pan bottom, you can end up with enough negative pressure in that area to pull through the oil that's above and suck in the air that's above the oil. The best pickup shape is similar to the best shape of an air entrance with a screen looking like about a third of a ball.

If you can't have the ideal setup, use a vendor with a good reputation to get started. Remember that stock pumps were designed to go 6,000 rpm. That was the end of their work. So some modifications such as stiffer relief springs and different sizing and routing of oil passages make these pumps work better. Most stock pumps will get the job done to 7,000 rpm with the bigger pickup and more pressure relief spring without anything else.

OIL SUMP CONSIDERATIONS

The oil pan on a wet sump has to be customized for the specific kind of racing you do. The pan size, shape, and baffling will vary with circle-track distance, shape and bank of the track, length of the straights, amount of lateral traction you get in reference to the banking, body shape, wings or no wings, track surface (dirt or pavement), tires, car, and track rules. The name of the game is to get the engine as low as possible in the car but have enough oil pan to keep the pickup under the oil 100 percent of the time and then try to keep the oil off the crank. Now some circle trackers have run sports car tracks, which have lateral acceleration (turns) in both directions. This is a real bitch. You've 'bout got to have a swinging pickup and an accumulator on this type of track because of the heavy acceleration, heavy braking, and lateral acceleration in both directions.

As far as changing anything on oil pans, if you do any welding on a steel pan, get the procedure right. One of the biggest problems with modified oil pans is cracked welds. Wet sumps might need some help, but dry-sump oil systems are pretty well worked out. Whichever system you are using, you need a specific pan design to fit various tracks and racing rules. In wet-sump and dry-sump racing, the deeper the oil pan, the better it works, so consider that as you choose your pan design. With a dry sump, if

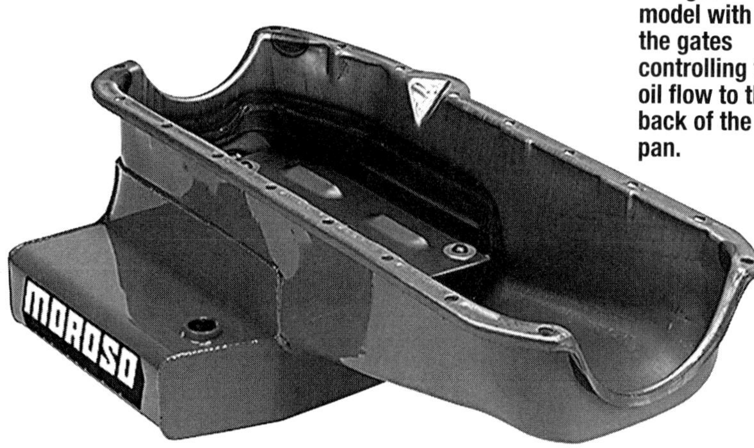

This Camaro racing pan is a five-gate model with the gates controlling the oil flow to the back of the pan.

Dry-sump units require the use of a tank, and these must be built with care. Every aspect of an oiling system is expected to perform at peak.

you under-scavenge for whatever reason, you get in a world of trouble. The crank gets partially submerged in oil and eats power while heating the oil really badly—all this will eventually break the crank. The best deal is to get a pump that can pull 3 to 5 inches of vacuum in the pan and keep it at a high rpm. That keeps the oil off the crank and reduces the air content in the oil.

Do all you can to pull a vacuum in the engine. It's best to run a sealed engine (no external vents) and plumb the dry sump so the oil sump tank does the venting for blow-by. Most sump tanks return scavenged oil to the top of the sump tank and have an arrangement of baffles throughout the tank. These baffles are arranged in a spiral configuration with the idea being that as the oil travels the spiral, the air in the oil will be removed. This flat don't work. Once the air is beat into the oil, it takes three to four hours for it to get separated. The best way to handle aerated oil is to return it directly back into the oil in the sump. This doesn't take air out, but it doesn't get worse falling down from the top of the tank through air to the oil reserve. A dry-sump pump with an air-separator section and oil that is being returned to the oil in the sump, along with a mesh filter, will keep 90 percent of the air out of the oil, but this only works at high rpm. The air that gets trapped in the oil on cold start and low rpm is gonna stay in the oil. All this assumes you are running a sealed engine.

In the last 30 years, there have been a thousand ideas tried to scavenge oil and keep the oil off the crank. The best deal is a directional, diamond-plated windage tray as close to the crank as possible with a vacuum being pulled in the oil pan. For those who run dry sump, a cogged beltdrive is best, running the oil pump about half the engine speed, but be careful about protecting the belt from track debris. A favorite failure is a ruptured or thrown belt from track dirt.

PULLEY PROBLEMS

For you high-rpm cats up to 8,000 or 9,000 rpm, watch out for all those pulleys and junk you piled on the nose of the crank coming loose. In the first place, the nose of the Chevy crank needs another 4 inches on it for a dry sump. Then, all the drive parts would have the same inside diameter, use the same keyway, and the fastening bolt would be short and simple. This way, the bolt could be in tension without fighting a jumble of unbalanced parts and the harmonic balancer (which is feeding torsional frequencies into the forward parts along with 'bout three different belts and their vibrations).

I think the front balancer and all that's in front of it should be put on a mandrel and all balanced together. If you run a fan, it should be balanced as well as the water pump and any other pulley (which includes the alternator). How many of you have broken an alternator housing? There is 'bout 30 to 35 hp feeding off that stack of pulleys.

We still have a lot more to learn. Clearances and oil hole sizing along with the overall routing of the oil through the engine will be explained. Also, topics like optimum oil temperature and the venting of the oiling system, what basics to follow when troubleshooting an oiling problem, how to examine failed engine remains and figure out a fix, and how to determine what products you don't need will be discussed.

Before we move on to the new topics, I would like to clarify one point I meant to mention. Concerning synthetic oil, another advantage to synthetics is that they can be thoroughly recycled after use. They don't wear out. The synthetics lose their additive package and pick up fuel and water from the heat and shearing action in the engine, making them heavier, but the oil doesn't break down. This protects the engine from failure better than mineral oil and helps to reduce waste. Now, let's move on.

CRANK CROSS-DRILLING

Because the stock block was not designed for racing, as time passed, racers made changes to its oil routing and passage sizes; they plugged some oil

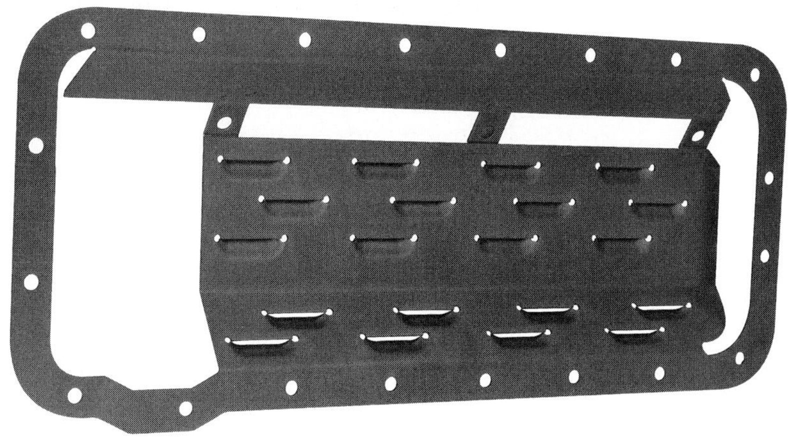

A well-designed windage tray will help keep oil off the crankshaft.

passages or restricted them, closed casting holes, and made other modifications. Some of these changes helped, some were a waste of time, and some hurt. Changes like cross-drilling the crank work because this is how the cam, rod, and main bearings are provided with oil to form the oil wedge at the proper timing. All operate in a full 360-degree arc at high speed with pressurized oil entering from a common point in the bearing. All three (cam, rod, and main) types of bearings see varying loads in their rotations. If you had a shaft at rest in a bearing with seals on both sides of the bearing and it had, say, 0.002 inch clearance when you pressurized it with oil, it would lift off of contact at the bottom and be dead center in the bearing. Well, even 100 psi of oil pressure in a running engine wouldn't be enough to carry the firing loads on the main and rod bearings, but we got a big goody working for us for free. In the rotation of the journals, an oil wedge is formed. The combination of the pressurized oil from the pump and the hydrostatic wedge can support all the loads in the engine except detonation.

When tearing down a failed engine, have you ever noticed rods and mains that are cracked, burnt, scoured, or chipped away? Detonation puts so much pressure on the piston that it overcomes the hydraulic support, and you have metal-to-metal contact. Cross-drilling is a good idea because it makes the lube job work. The position (timing) of the rod journal oiling hole in the big end of the rod (in reference to the piston location) should be placed properly to handle the varying loads, considering compression, firing, and return load when the piston is stopped at TDC and BDC. The changes in direction are a big load by themselves, considering the multiplication of weight in reference to rpm. Cross-drilling is one of the good ideas racers have come up with.

WRISTPIN OILING

A bearing surface that has been ignored is at the wristpin. Whether it's a floating or pressed-in pin doesn't matter, you have just three bearing surfaces doin' the same job as two bearing surfaces. You would think that three bearing surfaces would be better, but in real life, the combination of the wristpin turning in the rod and in the piston gets the rod bushing in trouble regarding heat from combustion and petrochemical changes. This is especially a problem in a long-duty (long-term) solution. So for low-buckers, the pressed-in pin is probably the best way to go. For high-buck racers who are using bushed rods, you might as well start using rods with oil holes (center-drilled rods). That bushing needs oil pressure of some kind to stand the rpm and horsepower loads of today's and tomorrow's engines. On wristpin clearance for either plan, it's 'bout the same. You need at least 0.0010 to 0.0015 inch. Wristpins are designed to be oiled by splash and the "Good Fairy"—yesterday's engines had rods and mains lubed the same way . . . splash and hope.

ELIMINATE OIL BYPASS

Changes to the oil passages in an engine block don't make a difference if you don't have properly machined—which means straight and untwisted at right angles—rods, cranks, cylinder bores, and other critical pieces. Nothin' will work if the pieces don't fit right. It's the same with detonation. The oiling system can't overcome these problems.

I've seen a hundred plans for changing the oil homes in an engine block for improved oiling. Some try bigger holes, plugging the lifter galleries, grooving the main and cam bearing areas in the block, fully grooved bearings, half-grooved bearings, increasing the oil hole size in the bearings, and other ideas. If you are a low-buck racer or a pro using a stock-type engine, the first action you should take is to eliminate the oil-bypass feature because it can open as low as 30 psi and dump a lot

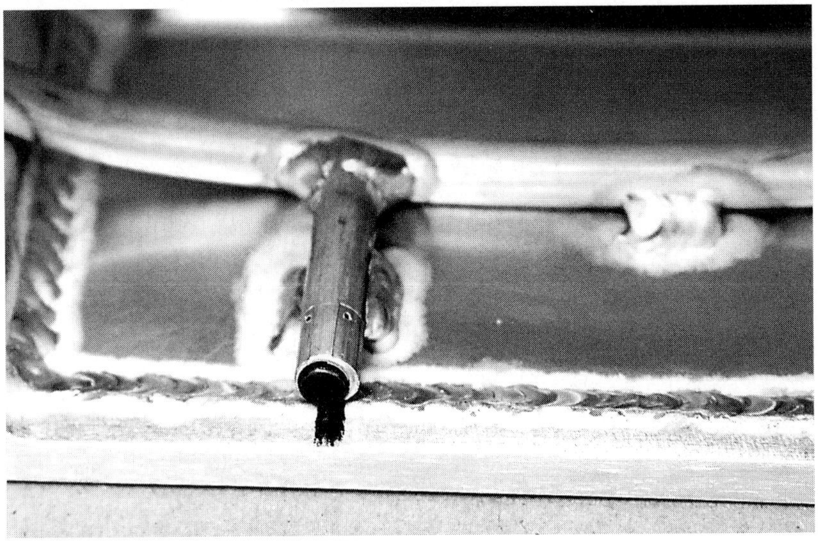

One way to get the oil going in the right direction is through the use of oil squirters in the valve covers. The oil is pumped through the aluminum tubing and squirts out in a fine spray through the holes near the rail.

more oil back into the sump. Car manufacturers have that on the engine to combat a plugged oil filter and to stop the oil filter from swelling and leaking on the fast-rpm cold starts, when the pressure could jump to 100 psi for a tenth of a second. You can get racing oil filters now that can handle 300 to 400 psi, so don't worry about the filter plugging up the works—get rid of the bypass.

There's no way we can describe what oiling schematic to follow on all engines used in racing today, so I'm just going to use a small-block Chevy as an example. If you didn't do one damn thing to any of the oiling system ('cept remove the bypass), including not restricting the lifter gallery oil flow, you could make 500 hp and turn 7,000 rpm.

OIL GALLERY RESTRICTION

Let's look at restricting the oil gallery flow. Most engine builders go down around 0.100 inch for a restrictor jet, juggle that 0.010 or 0.015 inch to the left or right, and then add a spray bar or flood the valvesprings to cool the springs. Should we use restrictors to reduce the amount of oil to the top of the head, and external systems to flood the valvesprings? You can see the plan is rather confused. I'd suggest restrictors from lifter galleries around 0.125 inch then don't do anything to the cylinder heads for low-buck racing. For high-buck racing, put standpipes at each end of the cylinder heads (drainback holes) that are high enough to flood the bottom two to three coils of the valvespring. Put a 0.090-inch hole in the bottom of the standpipes so when you pull the valve covers off you don't have a quart of oil run all over the exhaust system. Wait five minutes before you pull the valve covers. Use good valve seals on the intake and exhaust valves. Yup, those seals will hold the oil just fine.

The path the oil takes from the oil gallery to the top of the head is through the lifters and hollow pushrods. The hole in the top of the pushrod lines up with the hole in the rocker for a certain amount of time (degrees of crank rotation) to provide the correct amount of oil to the head. If the pushrod is too short or long, you screw up the timing of the holes and you can burn up the rockers. The problem of trying to cool overheated valvesprings is that it's hard to get enough oil up there.

With only so much oil flowing over a given amount of time, you need to determine where the oil is going. If you consider 16 lifters in a cast block leak quite a bit, that is a lot of oil pressure being bled off. It's the same with bearing clearances on the rod, main, and cam bearings. It used to be that 0.002 to 0.003 inch on rods and mains and 0.024 inch side clearance on the rods was what everybody ran, but now that is only used on engines that run to 7,000 rpm. Over 7,000 rpm, you cut the rod and main bearing clearance down to 0.00175 to 0.002 inch and the rod side clearances to about 0.0016 inch. Also, your oil gallery restrictor size affects oil pressure. Plugging casting holes in the lifter gallery is done by many people in the battle to reduce the amount of oil that drains down from the top of the engine onto the crankshaft, which costs power and heats and aerates the oil. This is a fact, but stopping that oil flow is a can of worms. The cam depends on splash oil—ain't that ridiculous? Here it is, 90 or so years into internal combustion engine development! If you stop the oil from getting to the crank, the problem is that you short the cam of oil. For low-buck racing, you best do nothing 'cept maybe epoxy a screen over the holes to catch any broken valvetrain parts. For high-buck racers, put standpipes on the right-side drain holes (numbers two, four, six, and eight) 'bout 1 1/2 inches tall. Just screen the left-side drain holes. But for everywhere there is a big hole bored in the lifter valley behind the cam gear for oil drainage (like on the small-block Chevy), put a freeze plug in on the right side (if you're looking at the engine from the rear). Don't bother with the hole on the left side. That right-side hole shoots an oil stream 1 inch in diameter right back to the distributor.

BLOCK MODS AND TIPS

There are a few things that should only be done

with the high-buck engines. It's a good idea to drill a small hole 0.025 to 0.030 inch from the front main to oil the cam thrust area. The timing chain gets way too much oil, so don't encourage it in that direction (on the high-buck jobs). Do what you can to reduce oil on the lower sprocket or the gear and chain. Now the holes in the Chevy small-block, especially on the center gallery in particular, should be redrilled. Chevy goes in from either end now and the intersection is off. It's not an easy process, but a 0.500-inch drill is long enough to go clear through, and that helps oiling. Even a 0.470-inch drill will get the job done. If you want to groove behind the cam bearing and mains, I don't think it does much but I don't think it hurts, either. I don't think increasing the hole size in the main or cam bearings is any help, but some of the very best engine men open up these holes. They have their theories and they win, but so do those who don't do this.

Stock engines usually have the oil flowing from the back to the front in the oil galleries. There are three theories that can be followed when making changes to this flow pattern. This oil can come from the rear, from the center, or from the front of the engine. In a Chevy, it is best to oil from the front. The center is not really doable. Actually, there is another choice. You could oil from the front and rear. I did that a lot. I would run a –6 line from the back to the front. For low-buckers, oil from the standard position. For high-buck deals, there is a 1/8-inch pipe hole drilled to the main oil supply center gallery. Put the block on a mill and carefully follow the hole angle. Open it to take a –8 fitting and send all your oil in here. Ninety-five percent of the Nextel Cup guys use this deal.

OIL PRESSURE AND TEMPERATURE

I still think 10 psi of oil for every 1,000 rpm is the hot oil pressure you should run. Some racers run a little more, but that's just lost horsepower. It takes 10 or more horsepower at 7,000 rpm to turn the oil pump. The more pressure you run, the more horsepower you lose. Some Nextel Cuppers drop the oil pressure to 50 psi when they qualify. I don't recommend that, but I guess a pole position is worth a lot of money, for various reasons.

Just as you need to maintain oil pressure, you must also maintain oil temperature. As for an oil cooler, you need to determine if that's what your engine needs. The oil should be run at 200 degrees F, at least, and 240 degrees F to 250 degrees F is the desirable upper limit. Mineral oil is in deep trouble at 290 degrees F because I believe the oil temperature at the rod bearing is 30 degrees F higher than any location we desire to measure.

Another way to keep oil temperature in check is through the use of coolers. Companies like B&M make plenty of options for cooling fluids in a race car.

Now, synthetic oil can run 380 degrees F in the sump and not be in trouble. But at those temperatures the bearings are in trouble. The oil cooler you choose must have enough oil passage surface area and line size so it doesn't restrict oil flow at all. It must have access to ambient air and a good exit path for the hot air as it passes out of the cooler back to the atmosphere. The area of the cooler transfer surface (fins) controls the amount of heat removed from the oil. You can get too small on a cooler, but it's hard to get too big.

An important part of the oiling is plumbing. You have to limit the line distance. Check the fittings for pressure drop. Avoid the 90-degree machined fittings—each one costs about 5 psi in pressure drop. Use long sweeping bends and a maximum of 45-degree tube fittings.

Once you know you're maintaining the temperature throughout the engine, the pressure needs to be checked at both ends of the system. You need 10 psi per 1,000 rpm with hot oil at the end of the system. With two gauges, you'll be able to think of ways to reduce the pressure drop, save some horsepower, and lower the oil temperature. The higher the pressure, the higher the rpm and the hotter the oil. When you have oil lines close to an exhaust heat source, insulate the lines.

OIL FILTER LOCATION

Regarding where to put the oil filter—you don't want dirt in the oil pump. You don't want dirt in the engine. You don't want dirt in the oil cooler. You don't want dirt in the sump tank. So, where are you supposed to put the filter? On a dry sump, my choice is to filter the oil coming back from the sump; then cool it by pulling with the suction side of the pump and then filter it on the pressure side of the pump so that it is clean going into the engine. Now, this setup ain't perfect. We don't want dirt in the cooler, and we don't want dirt through the pressure side of the pump— but most

Breathers attached to the valve cover are a common and efficient setup. This gives adequate crankcase ventilation while oil is pushed to the right by centrifugal force.

important—we don't want the engine to see any dirt. When you buy a cooler, you want a take-apart version you can clean in a sonic cleaner. You should use an oil pump (dry sump) with an iron pressure section and gears so if some dirt gets in there it can't grind up the gears and housing. Aluminum makes it worse.

In a wet sump, filter the oil before it goes into the oil cooler and back into the engine. Again, watch for a pressure drop from undersize fittings, too many 90-degree fittings, and small lines. Small filters and small coolers are bad, too. If you use an accumulator, make sure line size is not restrictive.

BREATHING

Engine oiling system breathing can be a mess, yet it is simple. In circle track racing, you turn left and oil in the right, so the rocker cover can fill up. If you have vents there, it can come out to a catch tank or out the breather caps. If your engine is in good shape, you can get all the venting you need on the left side only and the problem of oil out the breathers is not nearly as bad. What causes oil to blow out? If the rings didn't leak and we could seal the engine up as the pistons went up and down, no pressure would build. In real life, the rings do leak. They leak an average of 8 to 10 percent. So, if you are using 800 cubic feet of air a minute (cfm), there could be 80 cfm going into the pan. If there was no vent, pressure would build and air and oil would eventually get pushed out. Remember that the oil pan and rocker covers are all in the same cavity. So you not only have to handle the crankcase pressure but also the forces of gravity, acceleration, and deceleration. Left-side vents should handle it, but if you are detonating or the engine is leaking compression, you may need both left- and right-side vents on the valve covers and, in extreme cases, tie the left and right together with a tube. The better the lateral traction a race car has and the higher you wind the engine, the more centrifugal force and more blow-by you get. All this applies to wet-sump engines.

Dry-sump engines have the same problems but with more options to solve them. You can use the sump tank for a vent and you can add an auxiliary tank to catch it. Oil and condensing vapor get put back in the sump tank. You can scavenge directly from the right rocker area and/or oil gallery, removing both air and oil. As simple as this sounds, many racers make a few changes or shortcuts and have one hell of a time getting these systems to work. When you get involved in your first dry-sump system, buy the complete setup, read the instructions, and follow them to the letter. Once you get it working, you can try your inventions. If you goof, you can get back to safe. No matter how good you do everything, cleanliness is gonna make you or break you. Most people do a good job on a new engine and let the plumbing job get dirty. It all needs to be hospital clean. Most people get in trouble reworking an engine after a cam or bearing failure. Once the oil has dirt or metal run through it, you must go all the way down to totally clean on every part of the oil system. There are no shortcuts.

REBUILD AND DISASTER READING

When you pull an engine down for rework or after a major disaster, you have to operate like a detective working on a murder case. Examine everything to do with oiling. In the case of a rebuild, look for signs of trouble. If you find something, think out solutions. If it all looks great, write down all your clearances and assembly tricks in a notebook for future reference in case you get lost.

For a failed engine, it's damn hard to tell what was the cause. It's kinda like, which came first, the chicken or the egg? It could be (1) a broken rod with a good rod bearing and wristpin, (2) a broken rod and burnt rod bearing, or (3) a broken rod with a broken rod bolt and destroyed bearing. Number one is caused by a weak rod, number two from an oiling problem, and number three is a can of worms. Number three could be bad bolts, an oiling problem, or you don't know.

Of all failures, 70 percent of them are probably goofs on your part. Five percent could be a bad part or poor part selection. Fifteen percent could be an improper lubrication plan or detonation. Ten percent could be the use of the wrong lubricant. You have to be honest with yourself. Our nature is to blame anything but ourselves. Unfortunately, that approach leads to the same failures again. You can tell if a bearing is burnt, cracked, or chipped or if you have uneven or irregular wear on components. These situations indicate misaligned parts, overheating, wrong lubricant, detonation, fuel dilution, water intrusion, dirt in the oil, insufficient crush on the bearings, poor machine work, or inferior parts.

In my entire racing life, there has been a never-ending barrage of new miracle additives and new miracle lubes. At best, 1 percent of them help. This is not to say none of them are any good, but to caution you about looking for horsepower. The valvetrain is the biggest mechanical power eater. Rings are nowhere near what you think they eat in power, and bearings even less. A good test for these miracles and friction reducers would be in Formula 1, Indy Car, Nextel Cuppers, Sprinters, and Midgets. So who uses it? In most cases, it don't hurt, but some do. Some help, but 5 hp or 1 percent improvement would be a fair estimate. Today, it seems there are more of these additives than ever, and some may have possibilities, but mostly in reliability—'course that in itself is important.

After all these suggestions, make sure you follow some basic startup rules. Don't start engines cold. Don't start engines without oil pressure. Don't believe that if you run the starter to get oil pressure then add ignition that you are home free. You don't ever want to start with cold oil and no pressure. There is an affordable solution. All you have to do is add a block heater or an oil tank heater. They are cheap and weigh next to nothing. Then add a disconnect to the pressure side of the oil system on the engine. Make a 1- or 2-gallon tank strong enough to hold 60 psi, safely. Pressurize the tank to 60 psi, hook the tank up to the disconnect long enough to get the engine to 40 psi on the engine oil pressure gauge, and start. This will keep you from hurting the engine on startup.

Chapter 20

WATER PUMPS

A Compromise Between Power Lost, Power Gained and Proper Cooling

Text and Illustrations by Jim McFarland

This disassembled Edelbrock coolant pump (#8810) reveals typical elements of a high performance or racing design. A curved, six-vane impeller configuration is the result of specific flow-rate and pressure requirements. Design objectives include optimizing pump output with minimum parasitic losses.

There are some specific similarities between how an engine's water pump functions and that of a dynamometer water brake. In particular, each operates in a liquid medium, causing energy to be transferred by the movement of a rotor. In the case of the dyno environment, a measurement of engine torque is determined from rotor activity. On an engine, a pump's rotor helps circulate coolant throughout coolant passages that typically includes a radiator. Before we investigate the particulars of racing coolant (water) pumps, let's examine some fundamental reasons for engine cooling.

WHY IS COOLING IMPORTANT?

There are many reasons. Of them, the prevention of detonation is of high priority, not only because of lost power, but also from results that usually lead to parts damage, which is often quite expensive. Lubricant breakdown, thermal stresses, thermal strains and premature parts fatigue are also associated with inadequate or improper cooling. Even damaged cylinder-head gaskets, distorted piston bores and lost cylinder pressure can be traced to a faulty or ineffective cooling system.

Furthermore, a considerable body of data shows heat rejected to an engine's cooling system (gasoline fueled) is more dependent upon rpm than load. This fact alone is worth considering when contemplating coolant pump selection and the rpm range in which a racing engine is to be operated as being more important than power level.

Premature combustion—By way of review, in a spark ignition engine, there are two fundamental conditions of combustion not initiated by timed, controlled spark: pre-ignition (occasionally called "surface" ignition) and detonation.

Pre-ignition originates from premature combustion begun by overly heated surfaces or components within the combustion space. Glowing bits of carbon, heated edges of valves, excessively hot spark plug tips or other high-temperature areas exposed to pre-combusted air/fuel mixtures can cause pre-ignition.

Detonation (or "self-ignition") could be called "spontaneous combustion," the result of excessive temperature and pressure conditions within the combustion space. Characterized by "knock," bits of material found on spark plug porcelain, lost power and damaged engine parts, detonation produces very rapid and high cylinder pressures in an uncontrolled and potentially damaging fashion.

In both cases, pre-ignition and detonation, the earlier in the compression stroke these occur the greater the amount of negative torque applied to the crankshaft. Also, both conditions can be aggravated by excessive engine temperature, again relating to overall cooling efficiency.

Cooling Affects Volumetric Efficiency—All other related factors aside, the cooling of inlet air and that contained within the combustion space at the completion of intake cycles tends to boost volumetric efficiency and power. The lower the temperature of inlet charges the higher their respective masses and the higher the engine's VE. Stated another way, cooler inlet charges tend to have increased oxygen content. The more oxygen available, the more potential power. And, because of the inherent relationship between VE and torque, the higher the volumetric efficiency the higher the potential torque output—yet another reason for maintaining control over engine temperature.

Mechanical Benefits From Proper Cooling—Engine parts

operated at temperatures beyond their design limits can hasten material fatigue and failure. The beneficial effects of lubrication can also decrease. Besides heat- and pressure-related damage from pre-ignition and detonation (already discussed), head-gasket rupture and cylinder-bore distortion accompany excessive operational temperature. Parts failures can also result from prolonged high-temperature exposure, causing excessive thermal stress from sustained or repeated heat cycles.

It's also noteworthy that not just high temperature can lead to problems.

Temperature differences (or "gradients") often found in the area lying between intake and exhaust valve seats are a common occurrence. Operationally, hot exhaust valves and cooler intake valves (and their respective seat areas) combine to produce a large temperature gradient, intake to exhaust. In particular, this is one region where insufficient overall engine temperature control can lead to specific damage.

Another area of high temperature concerns piston top ring lands. Sudden combustion pressure and temperature, typically attendant with detonation, can lead to damage in this location. In fact, if temperature in this area exceeds 200 degrees C it's possible for engine oil to lose its ability to properly lubricate, leading to ring land collapse or fracture.

BASIC WATER PUMP FUNCTION

Nothing mysterious here, but some areas of discussion would be beneficial. Keep in mind the greatest flow restriction experienced by an engine's coolant pump is often created by cylinder heads and gaskets.

Many stock water pumps are basically low-efficiency designs. This is one area where those intended for racing can provide distinct improvement. Depending upon shaft rpm, impeller configuration and cavity, water pumps can create parasitic losses to gross horsepower, increasing net power reductions as a function of pump speed and component design. In a very general sense, they're small power absorption units, not completely unlike what you would view by examining the internal components of a water brake dynamometer.

To reduce the potential for cavitation and lost pump efficiency, designs should include a study of how coolant flow requirements relate to parasitic losses. Therefore, as might be expected, a trade-off exists between power loss and cooling gained, involving adequate flow and engine rpm.

Compared to their stock counterparts, coolant pumps designed for racing often contain features derived from solving problems associated with cooling high-output engines.

This coolant pump dyno, designed and built by Edelbrock R&D staff, leads to the provision of specific data that includes pump shaft rpm, input power, efficiency, inlet and total pressure, total flow and total pressure. By this method design criteria and evaluation of competitive pumps is obtained.

For example, Evans Cooling Systems entered the market some time ago by first identifying problem areas relative to material improvement. Then they addressed functional concerns engine builders had faced for years, including pump flow rates and volumes associated with high-powered engines in multiple racing situations.

Among these, a full-time bleed hose connects the air bleed to the thermostat housing, enabling vapor and pressure to be exported into the radiator. Maintaining a pump's impeller cavity at capacity reduces or eliminates cavitation and efficiency losses. Evans also provides a plug for the weep hole in the pump housing, intended to prevent air from being drawn through the hole and causing seal damage.

Plus, according to Evans Cooling Systems VP & GM, Steve Pressley, "We've come to the conclusion that coolant pumps need to be 'tuned' to the rpm range of the engine. By this I mean each pump has a range of efficiency based on its design and shaft rpm, beyond which it simply 'flatlines' for any additional flow.

"Ours tend to be in the span of 4,500–5,000 shaft rpm. If an engine to which one of these pumps is installed consistently runs in the 7,000–8,000 rpm range, the pump's drive pulley ratio needs to be adjusted accordingly, in order to maintain shaft speed in the 4,500–6,500 range, certainly no less than 4,500. Correspondingly, if the engine speed range is below this span, pump-shaft speed needs to be increased. It's important to keep a pump operating in its shaft speed range of greatest efficiency."

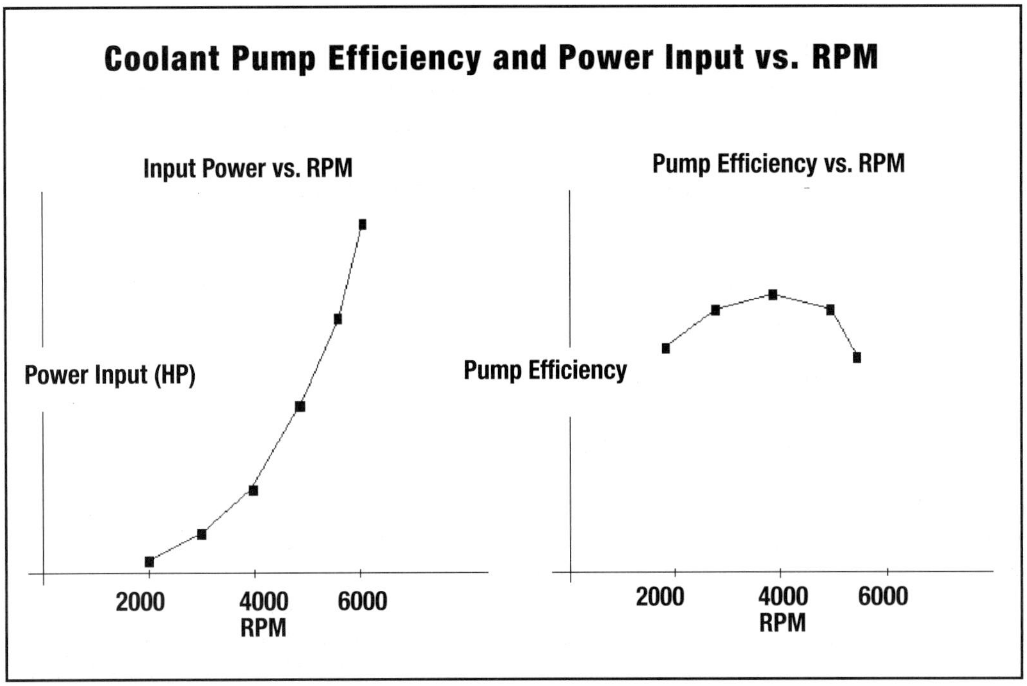

These two data charts provide emphasis for the suggestion coolant pumps be operated in their range of greatest efficiency. Note that at 4,000 rpm, less than half of the maximum power is required to produce peak coolant pump efficiency. Pumps should be "geared" with pulley ratios that enable shaft rpm to be maintained in the range of greatest pump efficiency, regardless of engine rpm. Courtesy Edelbrock Corp.

"FORWARD" AND "REVERSE" FLOW SYSTEMS

It is general practice that an engine's coolant enters the cylinder block from a comparatively low point, traveling up past and around cylinders and into the cylinder heads. We shall call this routing a "forward" direction. While coolant pump characteristics and design affect flow rates to different parts of the engine, it's also possible to regulate these rates by passage holes located in the cylinder head gasket(s).

One of the disadvantages of routing coolant in this fashion is that cooling of the combustion space is hampered by exposure to coolant already heated, having passed by outer portions of the cylinder walls. One of the problems this method of routing creates is an inability to mitigate or control detonation during prolonged combustion temperatures often produced in a racing engine, sometimes preventing optimized spark timing. Such spark limiting tends to reduce power potential in the interest of controlling detonation. Increased coolant circulation rate and proper residence time within the radiator cores also plays a role, thereby bringing coolant pump design and operation into play.

Simply going for "maximum coolant circulation rate" can lead to insufficient radiator residence time and excessive temperature. Coolant must stay in a radiator long enough to release heat into the atmosphere.

By changing this direction of coolant flow to enter an engine from the cylinder head area, circulating down and around cylinder bores, a "reverse" flow path can be created. While there are factors at issue that deal with system pressure and routing, a specific benefit of increased spark timing can often be used for a power gain, since combustion temperatures tend to be reduced.

By using this method, coolant delivered from the radiator is provided directly to the area in the engine with the greatest temperature: the cylinder heads and combustion chambers. Such a temperature gradient tends to increase heat removal, thereby providing some opportunity for increased spark timing, combustion efficiency, parts life, and power.

FACTORS AFFECTING WATER PUMP SELECTION

It's not just all about lightweight materials, parts durability, or the fact aluminum coolant pump housings have different heat transfer characteristics from cast iron. Pump impeller design, impeller shrouding, housing capacity, reduced or eliminated cavitation, pump speed, bank-to-bank flow balance (in V-type engines), and output volume are integrated into proper pump function.

Of the rules of thumb that can apply in pump selection, boosting flow in the lower- and mid-rpm ranges without significant losses in pressure and bank-to-bank balance is worthwhile.

In fact, according to *Circle Track* Technical Council member and longtime engine builder Keith Dorton, "We like to achieve about 50 gallons per minute output in the 7,500–8,000rpm range. In this regard, we've had particularly good luck with the Adams and Drake pumps, especially for the 'bullring' tracks where there's not much cool air—you're usually behind somebody, and there's a pretty strong reliance on the engine's cooling system, particularly the pump and radiator."

Depending upon the amount of power a racer is willing to sacrifice for added coolant flow, a loss of 2–3 hp may not be important. But to the degree contemporary engine builders are willing to search for consistent 2–3 hp gains, coolant pump selection may be a more critical issue. Therefore, if this is the case, it's worthwhile to dig a bit deeper into the pump performance specifications often provided by manufacturers.

ASK THE EXPERT

Rick Roberts is a graduate engineer with advanced degrees and has been spearheading Edelbrock's core research projects for the past 10 years. His experience and candor provide some valuable insight on the design, function and selection of racing coolant pumps.

CT: What do you consider the most important factors in the selection of a coolant pump for circle-track engine applications?

RR: "Simply put, water pumps produce flow and raise pressure at the expense of power to rotate the pump's shaft. A pump must provide adequate flow and pressure to maintain proper coolant temperature throughout all racing conditions, regardless of the race situation. This depends not only on the pump, but also on the remainder of the cooling system.

"In our pumps applicable to circle track racing, we've focused on reducing power consumption while still maintaining cooling requirements. I'd say these are critical factors in the pump selection process."

CT: How would you characterize pump impeller/cavity design as these apply to addressing an engine's cooling requirements?

RR: "The impeller should be sized to meet the flow and pressure requirements of the engine and cooling system. These are usually dictated by a combination of past experience and personal preference rather than detailed calculations, at least in our experience. Until recently, it has been assumed more flow with adequate pressure is the way to go.

"In an ideal situation, for example, let's say you designed an impeller/cavity combination that produced 20 percent more flow at the same pressure and shaft speed as before the change. In theory, that pump would require approximately 35 percent more power to turn its shaft. Depending upon circumstances, this increase could approach 1–2 hp of parasitic loss just to gain 20 percent more coolant flow.

"Today, many racers cannot give up 2 hp for an edge against overheating. Hence, there is now a need to consider all aspects of the cooling system, and in particular, the pump's power consumption."

CT: What are the most common mistakes made in proper coolant pump selection?

RR: "Most racers seem to focus on flow rate as a key performance characteristic. Not to imply this is a mistake, the fact is there's more to consider than flow. Pressures, power consumption, right-to-left balance, variation as a function rpm and flow stability are all important factors when selecting a coolant pump.

"For specific investigations of these components, we've constructed a coolant pump 'dyno' to test pumps through a wide range of operating conditions. Of late, our research has pointed the way for specific design refinements that should make coolant pump selection considerably simpler in the future."

CT: What are your recommendations for matching coolant pumps to specific engine displacements, rpm and power level?

RR: "Let's talk about power level. I asked one of our engineers working on the coolant pump project to produce a thermodynamic 'budget' for a typical performance engine. By this, I meant he needed to calculate a summary of how the energy released during combustion was being consumed, either in useful work (turning the crankshaft) or lost to such unavoidable variables as heat rejection to the coolant, work loss to friction, heat loss and other conditions.

"To his amazement, only about 25 percent of the energy released during combustion was actually converted to torque on the crankshaft, with the 'lion's share' of the loss reflected as heat in the cooling system. Simply put, the greater the power level the more heat rejection, typically to the coolant. Different bore/stroke combinations may change this allocation somewhat, but I doubt it is significant.

"Here's one cursory comment on the effects of rpm on pump function. Our pumps have the greatest power efficiency in the range from 3,000 to 5,000 pump-shaft rpm. Regardless of the situation and for best results, pulleys should be selected to place the pump speed in this range."

CT: Do you have any views on matching radiator size (capacity) to cooling pump pumping capacity?

RR: "At this point in time, I do not. From my previous comments, it should be evident we are now looking more closely at the entire cooling system which, of course, includes the radiator. Obviously, a radiator relies on airflow to carry heat away from the engine coolant. All other things being equal (such as air velocity through the radiator), its ability to remove coolant heat is proportional to the area exposed to the cooling air. This means more rows of cooling passage in the radiator and/or a larger height and width of radiator.

"We're contemplating connecting a radiator to our water pump dyno, which employs a heater that enables water temperature up to 200 degrees F. By so doing, we can explore the interactive aspects of the radiator and coolant pump."

Chapter 21

COOLING FOR POWER THEORY

The Balance Between Creating Combustion Heat and Regulating Its Effects is Key to Optimizing Power and Parts Life

Text and Illustrations by Jim McFarland

Editor's Note: This chapter is compiled as an assembly of information about internal combustion engines and their cooling systems. It neither advocates nor suggests one system or method as superior to another. Rather, it is intended to stimulate thoughts and point to peripheral areas involving cooling-system performance worth considering when optimizing race engine power.

For the moment, direct your attention to an engine's combustion space. It's the heart of this chapter and focal point for the production of heat (power). Ideally, we'd like to optimize the containment and conversion of this heat into useable crankshaft torque while minimizing heat lost to the cooling system. Clearly oversimplified, this is the platform on which the following paragraphs are built. Internal combustion engines, particularly those used for racing, are a complex network of compromises. The balance between maximizing heat and minimizing its loss to cooling is among these compromises.

IT'S NOT JUST ABOUT COOLING HOT PARTS

Overheated lubricants cannot provide proper lubrication. Reasons include both chemical and physical breakdown of oil and a reduction of film strength. Results from this condition include thermally overstressed parts (particularly pistons), increased losses from friction horsepower, and mechanical failures. In the process, higher piston crown temperatures can lead to either pre-ignition, detonation, or both. Although pre-ignition and detonation event periods may be brief when compared to normal combustion cycles, additional heat can result from these conditions, too.

Torque output, a function of volumetric efficiency, can also suffer. This comes from increased air/fuel charge temperature, caused by what is called the "heat transfer effect" whereby heat is added to otherwise cooler mixtures, decreasing mixture density and reducing net volumetric efficiency. Conversely, a reduction in charge temperature tends to increase density and improve volumetric efficiency, suggesting a cooling system (or inlet manifold) design that routes heated coolant away from the inlet path. An example of this can be found in so-called "air gap" intake manifolds or by the elimination of a coolant crossover passage in V-type engines. Short of lowering engine mounting locations, this also provides for a lower overall engine profile for increased frontal area and improved race car aerodynamics.

HEAT TRANSFER AND THE NEED FOR COOLING

Arguably, most combustion heat created in a piston crown travels into cylinder walls by way of the piston rings. Unless proper cylinder wall cooling is provided, an abnormal wall temperature will reverse this path and lead to overheated piston crowns. Even if pre-ignition or detonation are not induced, the material strength of pistons may come into question, increasing the importance of cylinder wall cooling that is adequate but not a "sink" for reducing needed combustion heat (power). Again, a necessary balance between heat containment and transfer rears its head.

Cylinder head temperature control is also vital to optimizing power. While it is generally desirable to keep maximum head temperature below 400–450 degrees F, failure to do so can lead to overheated exhaust valves and spark plugs, either of which can produce pre-ignition, detonation, and damage.

Furthermore, it has been determined that even though combustion and the hot gases it produces are high in temperature, these gases are not good radiators. As a result, a process called "convection" creates major cylinder wall temperature or the transfer of heat from hot combustion gas-to-wall material by motion of the gases themselves. While this process may appear of little consequence to the importance of proper cooling, it is the primary means by which increases in horsepower require improved cooling efficiency, whether by coolant path, system design, coolant choice, or a combination of these.

ENGINE CONDITIONS AFFECTING COOLING SYSTEM REQUIREMENTS

Specific engine heat output can be affected by various operating factors. In one way or another, each of the following issues can have an impact on cooling system requirements.

Mechanical Compression Ratio—Because of a greater expansion of combustion gases near and at TDC piston position during the firing cycle, an increase in compression ratio will boost cylinder wall temperature during this time. However, near BDC on the power stroke, when an increased amount of cylinder wall surface area is exposed to the combustion gas, there is typically a reduction in its temperature (EGT) during early stages of the exhaust cycle (blowdown).

Any increase in cooling temperature (heat rejected to this system) will occur at or near TDC during combustion, thereby increasing an engine's sensitivity to detonation, spark timing, and cooling system efficiency. In terms of cooling system design, if a means were to exist that provided increased uniformity and control of temperature in and around the combustion area, an opportunity could arise that allowed higher mechanical compression ratios to be used without incurring abnormal combustion (note the sections on normal- and reverse-flow cooling systems).

Air/Fuel Charge Ratio—Both combustion flame speed and combustion temperatures fluctuate with changes in air/fuel ratio. Generally, a ratio of about 13:1 will produce the highest combustion gas temperature, therefore affecting the differential between the combustion space and cooling system adjacent to this area. At air/fuel ratios somewhat richer than this, maximum heat rejection to the cooling system tends to occur earlier in the burn and with less cylinder wall exposure. An accepted reason for this is that slower flame speeds associated with leaner air/fuel mixtures tend to expose increased amounts of cylinder wall surface over which to "spread" the burn during later stages of the combustion process.

Spark Advance—If ignition spark is used as a "crutch" to overcome other deficiencies in an engine's combustion process, it is possible to create additional heat. For example, given poor mixture quality (homogeneity), additional spark timing may be required to adequately burn the "leaner" portions of an air/fuel charge. This tends to add pre-TDC work on the piston, thereby increasing combustion temperature and boosting the cooling requirement. Correspondingly, if spark timing is too retarded, cylinder pressure and gas expansion beyond TDC will add heat to the cooling system.

Power Output—As either power or rpm (or both) increase, all other factors being equal, additional heat will be rejected to the cooling system. As piston displacement increases, the amount of combustion surface area increases accordingly, thereby providing more surface area over which to distribute combustion temperature. Therefore, "large" engines tend to lose less heat per unit area to cylinder walls than those of smaller displacement.

DEMANDS PLACED ON COOLING SYSTEMS

Discussion of how the foregoing factors affect engine heat is necessary to appreciate the benefits of cooling system design. Generally, of the total amount of heat produced in the combustion process, about 50 percent flows into the cylinder head and valve seats, 30 percent into the cylinder walls, and 20 percent into the exhaust port. Heating of intake ports can be largely identified as coming from heated coolant. More specifically, about 50 percent of the heat rejected to the cooling system is created during the compression, combustion, and expansion phases of the overall combustion process. In many instances, a higher overall percentage than this is lost when comparing the potential energy available from a given air/fuel charge and what finally becomes useable crankshaft torque.

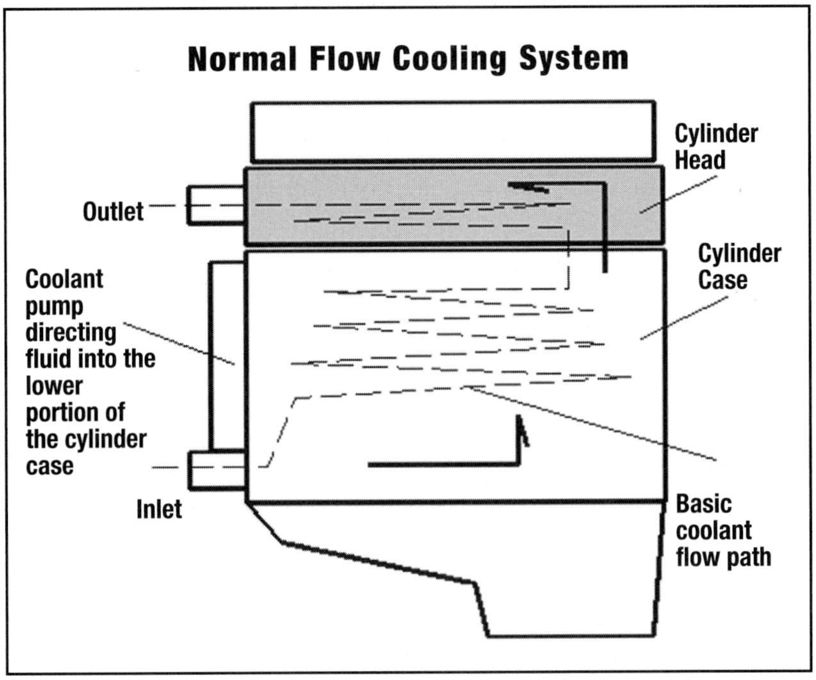

This type of system typically introduces coolant in the cylinder case first, circulating it upward into the cylinder head. In this fashion, cylinder head coolant is "preheated" by having passed through the cylinder case containing cylinders heated by combustion and friction.

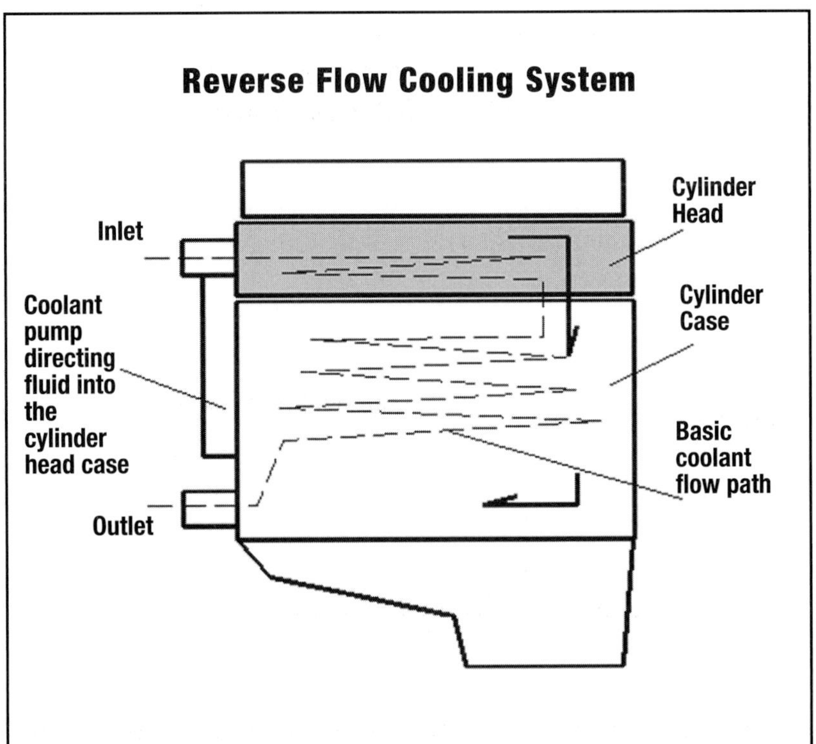

This type of system typically introduces coolant first in the cylinder head, circulating it downward into the cylinder case. In this fashion, cylinder head coolant enters the hottest part of the engine first before passing downward through the cylinder block and back into the radiator.

Whether cylinders are cooled before cylinder heads or heads before cylinders, there can be effects on engine performance and options available for optimizing combustion efficiency and power. In this regard, it may be helpful to also be aware that combustion flame control and power output is influenced by environmental temperature in the combustion space. In other words, the hotter the environment, the greater the tendency of (and compensation for) abnormal combustion and the problems encountered when trying to optimize engine performance.

"NORMAL" FLOW COOLING SYSTEMS

Common practice is to introduce engine coolant from the lower part of a radiator or heat exchanger into the cylinder block's jacket, upward into the cylinder head(s), and back into the upper part of the radiator. A thermostat is used to cause coolant retention in the engine until a pre-determined temperature is reached, at which time flow throughout this system is allowed, circulated by a pump. This is basic stuff.

Coolant in contact with the inner surfaces of coolant passages is directly exposed to heated metal surfaces. Depending upon the boiling point of the coolant and system pressure, "boiling" or bubble formation will occur at this interface. In fact, this event is evidence that the coolant is involved in a process called "nucleate boiling" and absorbing heat. Under certain circumstances, excessive boiling can lead to the formation of "steam pockets" that tend to separate coolant and heated surfaces, causing local concentrations of overheated metal and further problems (failed parts, abnormal or uncontrolled combustion, and lost power). So coolant boiling point, system pressure, and flow rate are among features that combine to efficiently absorb, remove, and control combustion- and friction-generated heat.

Regardless, this direction of flow tends to bring "preheated" coolant to locations near the combustion area. One condition created is a tendency to raise the temperature of the combustion space. This can lead to the heating of air/fuel charges prior to their combustion, thus limiting the amount of spark timing that can be used short of abnormal combustion. It also tends to place limits on mechanical compression ratio.

As a result, particularly in light of contemporary airflow technology and combustion techniques, such "tools" as specific mixture motion in the inlet path and combustion space have combined to improve burn efficiency, in part intended to overcome certain normal coolant flow problems.

This has resulted in various improvements in engine output, whether directed to increased power (torque), on-track fuel economy, or parts durability. It has also given rise to exploration to identify and refine coolants other than water, particularly those with higher boiling points.

"REVERSE" FLOW COOLING SYSTEMS

It is known to the art that introducing coolant to the area of greatest engine heat (typically the combustion space) before passing it into cylinder block jackets ("top-down" flow) offers the potential for improved combustion efficiency and/or power. Among the benefits is the opportunity to revisit ignition spark timing parameters, notably the possibility of increased spark timing and power, absent abnormal combustion. However, there have traditionally been problems associated with the introduction of coolant in a "downward" direction (e.g., from the cylinder head area downward into the cylinder block). One difficulty has been how to deal with formation of vapor and pressures in the upper part of the cooling system. The solution to this problem, given proper vapor venting, has allowed the reverse-flow cooling technique to become an "enabler" for companion engine modifications previously suggested.

In particular, it's worthwhile to understand some of the events that take place within an engine's cooling system in order to further appreciate the potential of reverse flow. For example, the higher the coolant temperature in proximity to the combustion space, the greater the possibility for steam pockets to develop. Steam pockets become "insulators" to the transfer of combustion heat into the coolant. The presence of these pockets, often located around the exhaust side of combustion chambers, can make a bad problem worse . . . frequently leading to abnormal combustion and damage.

For normal flow systems ("bottom-up" flow), numerous areas have been explored to resolve the problems associated with steam pocket development and resulting power limitations. These include "mapping" of coolant flow rates, flow volume, and distribution patterns. Juggling of coolant transfer holes in cylinder head gaskets has also been employed, along with special coolant pump features, enhanced radiator design, and chemical blends intended to raise coolant boiling point. But the fact remains—when coolant is delivered to the combustion area following a measure of pre-heating in the cylinder block, certain efficiency disadvantages are created.

What's the message here? To optimize engine

Modified Reverse Flow Cooling System

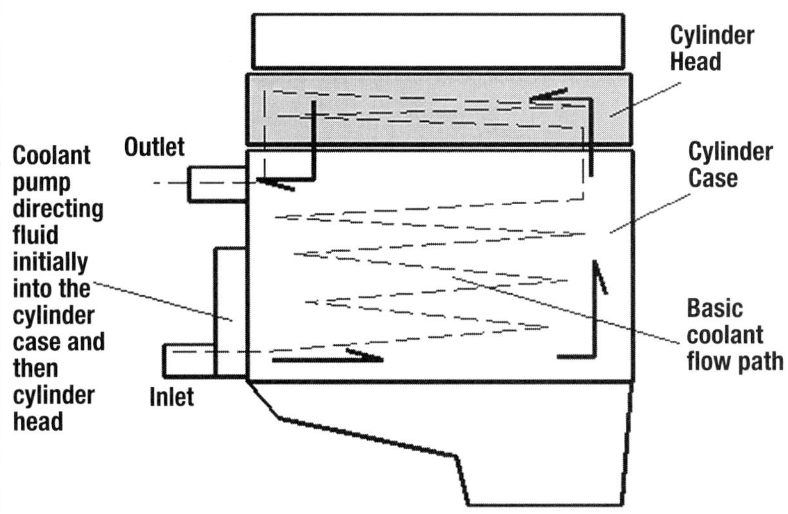

This type of system typically introduces coolant first into the cylinder case, then into the cylinder head, subsequently circulating it downward back into the cylinder case and out. In this fashion, cylinder head coolant enters the cylinder case and is routed to the hottest part of the engine before passing downward through the cylinder block and back into the radiator to prevent the flow of hot coolant away from and reduce its effects upon the intake system.

Changes in Heat Rejection As a Function of Air/Fuel Ratio

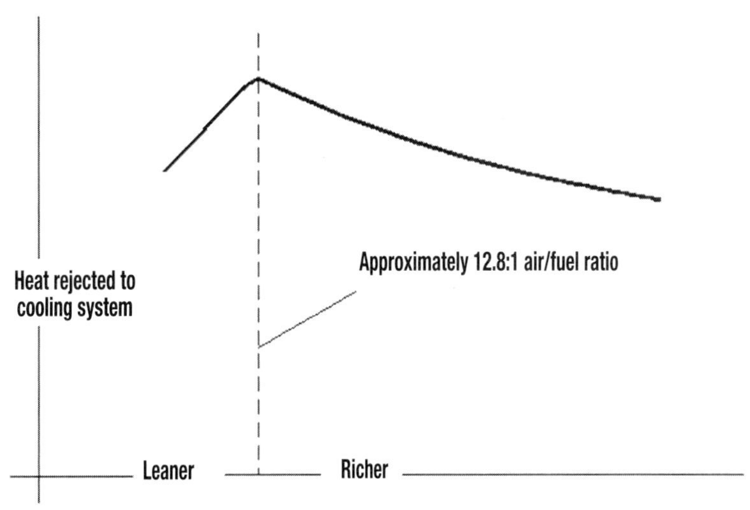

The purpose of this graph is to show how enrichment affects the amount of combustion heat rejected to the cooling system. As mixtures become richer, flame speed begins to increase, subjecting less combustion surface area to heat. Correspondingly, as mixtures become leaner, flame speed decreases, but so does the amount of combustion heat produced that must be addressed by the cooling system.

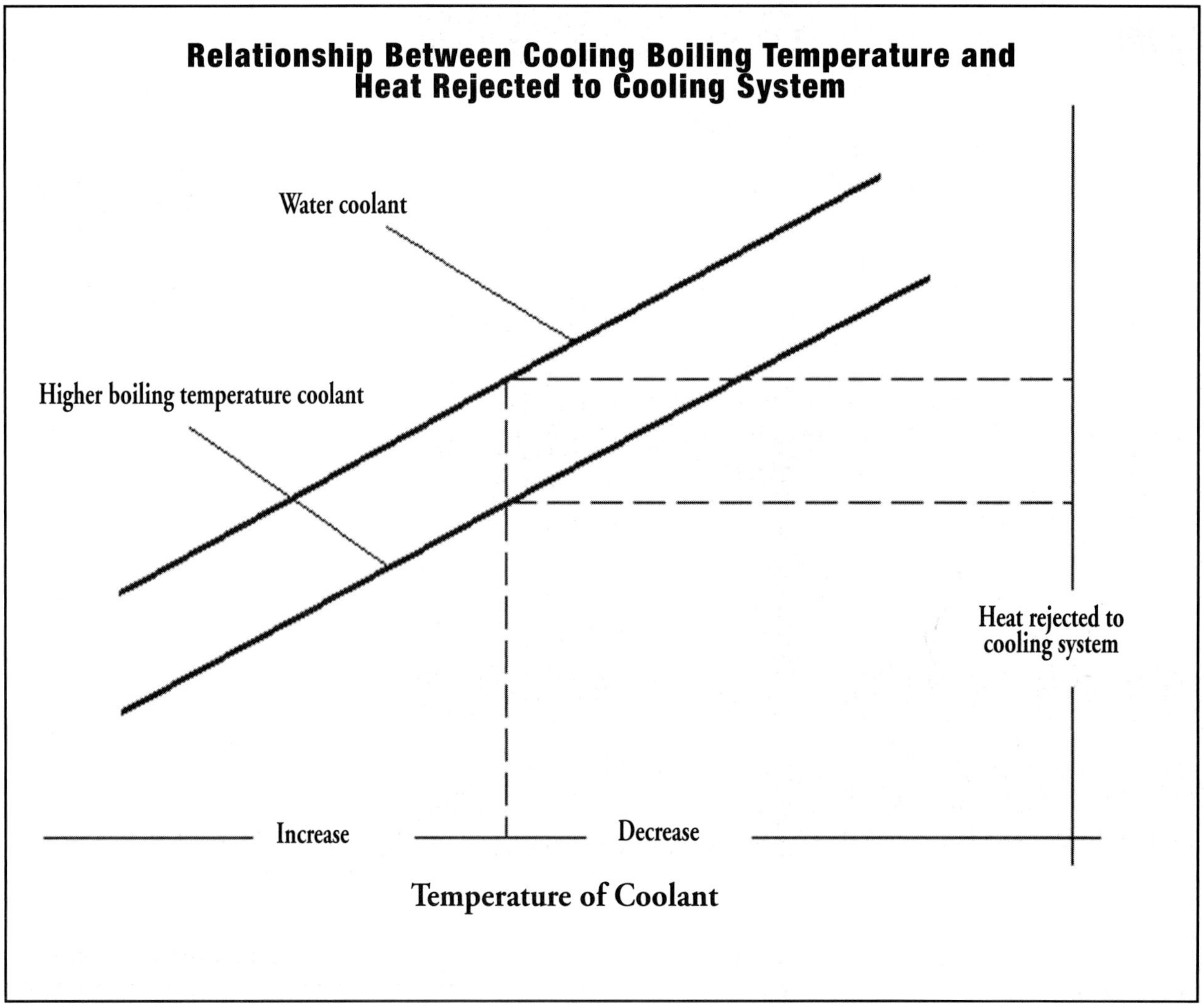

As coolant boiling temperature is increased by choice of coolant composition, the amount of heat rejected to the cooling system decreases. In other words, more combustion heat is available for conversion into power than with lower boiling point coolants.

power, a balance should be created between maximizing combustion heat and minimizing its adverse thermal and mechanical effects on parts. Short of abnormal combustion (primarily detonation), building and containing as much heat as possible within the combustion space can lead to increased horsepower. Of its jobs, the cooling system (including its contents) must help maintain combustion space temperature at a level that provides maximum heat release (power) during combustion without allowing the process to degrade into an uncontrolled burn. If this is accomplished, mechanical damage to engine parts sensitive to heat absorption will be minimized, thus extending the life span of these parts.

As previously suggested, an engine's components are a blend of functional compromises. Cooling system capability should be matched to engine characteristics and output. By installing a larger radiator, increasing system pressure, providing a greater volume of cooling air, or finding a high-output coolant pump, a given problem may be avoided or solved. It is likely more important to thoroughly understand their function and view cooling systems as "enablers" to optimizing power and extending parts life, treating them accordingly.

Chapter 22

AIRFLOW PATTERNS

Where Air Goes, or Should

Text and Illustrations by Jim McFarland

A simple dye spray may be enough to unlock more power in a racing engine.

Consider this: Engines don't burn just fuel, they burn fuel and air. The more air, the more power, and fuel delivery is adjusted accordingly…or so the axiom goes. In reality, air quantity must be equated with air quality, for without proper conditioning of mixtures, combustion efficiency will suffer. More air isn't always more power…as you will soon discover.

Some fundamental considerations Let's say you have an engine that produces a given level of power, demonstrates certain brake specific fuel consumption (BSFC), and requires a known quantity of spark timing. Further, assume this engine to be our "baseline" for discussion. As you might assume, these conditions are produced with what we'll also call "baseline combustion efficiency."

Now, without changing net airflow into the engine, suppose we perform a set of modifications intended only to improve mixture quality (homogeneity). In conventional terms, this simply means keeping more fuel in suspension prior to and during combustion. We want to emphasize that net fuel and airflow are not changed, only mixture quality.

From these types of modifications, some interesting results can occur. First, improved mixing (homogenization) tends to reduce fuel particle size. The smaller their size, the faster the burn, generally speaking. Then cylinder pressure rises more quickly, enabling slight reductions in total ignition spark timing. As either or both of these conditions develop, net torque on the crankshaft increases, largely by the reduction in negative work (pressure on the piston during combustion and prior to TDC compression position) and a quicker burn.

These conditions trace directly to improved mixture quality, which then becomes the enabler for torque increases. More to the point, it is the "construction" of improved air quality that supports more efficient fuel atomization and burn efficiency gains.

THE PROBLEM OF AIR/FUEL SEPARATION

This is a classical problem when changing the direction or flow rate (or both) of fuel-suspended airflow. The relative differences in compressibility and weight (by volume) of air and fuel leads to their departure during changes in energy state (flow direction, rate, or both). It is the condition of the air that tends to affect how fuel is delivered. As an analogy, the air stream could be equated to a flowing river and fuel likened to leaves thrown into the stream. The motion of the river stream significantly affects the direction and movement of the leaves. By affecting the characteristics of the stream, movement of the leaves can be directed. Such is the case when transmitting fuel particles through an engine's intake track.

Interestingly, many cylinder head modifiers spend inordinate amounts of time sorting out port designs, valve seat and bowl dimensions, and combustion chamber shapes

Short bursts of aerosol machinist dye work well to establish airflow patterns. Excessive use can mask the contrast of how the dye represents the intensity of where separated fuel will migrate. Before concluding inlet port and chamber designs, effects of the intake manifold should be determined. While this does not include the influence of a carburetor, it does reveal the impact of ports and runners working together. Note proximity of can to entry, as demonstrated by Wells Racing's Ricky Stapleton.

to optimize net airflow. Not to diminish the value of this approach to increasing an engine's output, especially if fuel is admitted up-stream of the combustion space, but omission of how such airflow affects the delivery of fuel can lead to an imbalance between increased air and power. Initially, the idea of using some form of atomized spray that would leave its mark on flow surfaces was conceived to address the air/fuel separation problem.

In time, as the windows of information about this approach began to open, it became evident that other aspects of improving combustion efficiency could be explored. In fact, this sequence of investigation led to the notion of using kinetic energy already contained in the inlet air stream to aid re-suspension of separated fuel by the method of dimpling or a comparable approach to surface imperfections. There are, of course, more sophisticated methods to study active flow and the influence of variables that affect its behavior and characteristics. However, most of you don't have access to CFD (Computational Fluid Dynamics) or comparable techniques. So at the more grassroots level, the approach described in this chapter is worthwhile.

One of the purposes of this chapter is to reveal the process of using both dye traces and flow path surface roughness to aid fuel atomization and suspension in the air stream. Although considered by some engine builders to be a "crutch" or "band-aid" to resolving a problem inherent in virtually any internal combustion engine, the latter of these two methods is proven, recognized by the U.S. Patent office as viable intellectual property, and

In this LS1 chamber, note the wash of fuel against the bowl area and across the spark plug boss. While this direction of mixture motion toward the exhaust side of the chamber is desirable, the heavy concentration of dye beneath the valve seat and absence around the backside of the chamber wall is not.

increasingly practiced wide-scale by builders of several types of engines. As previously stated, it is not "snake oil" and is worthy of the time taken to understand and practice, at least in the opinion of this writer.

In the real world of making power . . . Dennis Wells of Wells Racing Engines shared comments

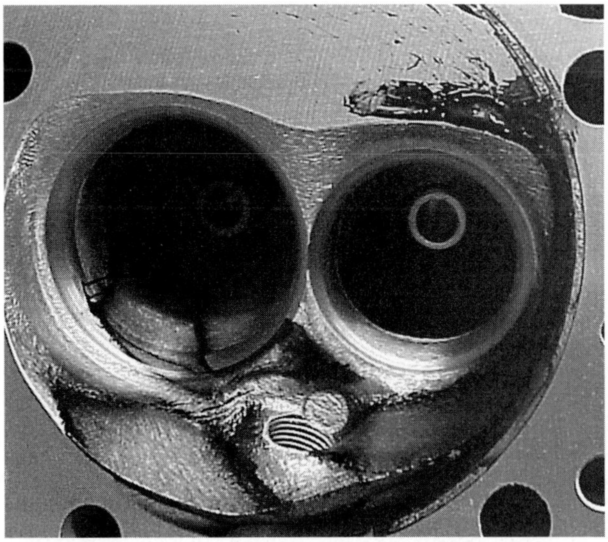

While canting the back wall of this combustion chamber may have improved raw airflow (as measured on a bench), note the concentration of fuel wash and lack of movement in the area of the spark plug nose. Rotation of the flow in a counterclockwise direction (shown) is favorable, but the collection of raw fuel just past the exhaust valve pocket is not. Matching such patterns with piston crown shapes and surface texture is important to boosting combustion efficiency.

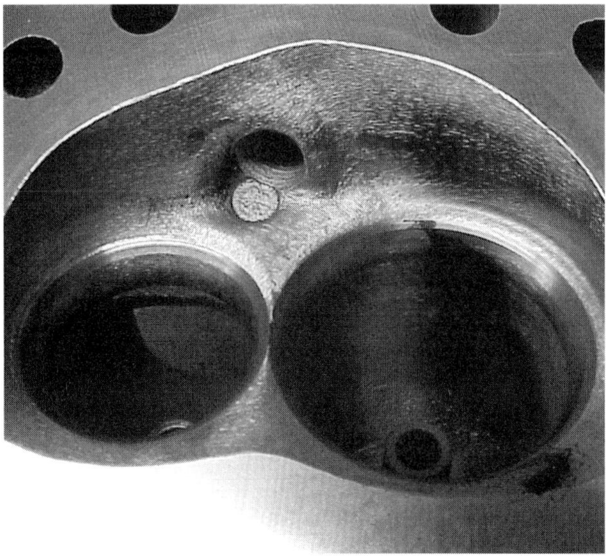

This is good. Note uniformity of pattern that includes activity in proximity of the spark plug nose and lack of "fuel" buildup, both against the chamber's back wall and sidewall of the exhaust valve pocket. Uniformity of pattern suggests a narrow range of air/fuel ratios will be present in the combustion space, as opposed to a wider range of ratios when patterns indicate sharper contrast of dye residue.

In areas where fuel is impacting combustion surfaces and "washing" into rich air/fuel ratio development, surfaces can be dimpled to create eddies or boundary layer activity intended to restore separated fuel into the air stream. While this approach may be considered a "band-aid" method, it can be highly effective, cost-effective, and power producing.

about his experience with the airflow/dye method of flow analysis.

"Before I began practicing this method, we'd tear down an engine, look at the burn patterns on piston tops and combustion chambers and wonder about what we saw. Sometimes, we'd see no residue; other times, we'd see varying shades of carbon buildup. Even though it occurred to us these patterns were probably an indication of something we should understand, we really didn't.

"Then, I was introduced to using a spray-type dye during airflow bench testing of cylinder heads. By studying areas where the dye was really concentrated, it appeared these locations indicated fuel wash or separation from the airflow. What really got my attention was when we started changing the surface texture of these areas and began to see power increases with little or no changes in total air or fuel flow. Exhaust gas temperatures also began to drop, suggesting the burn was taking place faster and that more of the fuel already in the combustion chamber was now being burned. We also discovered less spark timing was required for best power," Wells said.

"As I've used this method of airflow study, we've found other areas for modifications that help increase horsepower. One of these has been in working with two-plane manifolds and the use of carburetor spacers. I never realized why spacers seemed to make more power until we did some dye work and discovered the problem of fuel separating on the manifold's upper floor. We discovered that as the carburetor is raised, this separation problem was reduced and more power was produced from about the same amount of fuel and air we used before adding the spacer. In some cases, we found it was possible to lean down the carburetor and make even more power. Sometimes this wasn't necessary because raising the carburetor also tends to reduce fuel-metering signal and makes the engine leaner, anyway.

"Overall, there are several (construction techniques) we've found helpful using the dye method," Wells continued. "One is that it points to where flow surface texture changes will help. We like to use dimples in these areas because they tend to put separated fuel back into the air stream. What this does is allow otherwise wasted fuel to be burned. This is power, any way you look at it. We also discovered that how air is delivered to the combustion chamber, when measuring flow patterns with a head alone, usually changes when you bolt on the intake manifold. So just studying these patterns with the effects of the manifold can be misleading. We've worked on optimizing flow patterns in the combustion chamber and discovered all this changes when the manifold is installed.

"What's really interesting is that we've used this method to identify areas of fuel separation, created dimple patterns, and found power gains without

As a companion to textured combustion chamber surfaces, the combustion chamber "floor" piston crowns should be configured to further aid mixture quality, both prior to and during the burn. Location and arrangement of dimples is a function of burn and/or dye patterns, as outlined in the text.

any significant changes in the amount of air flowed. Plus, we've also discovered that what we'd previously thought were helpful modifications in an intake manifold that increased air flow actually caused separation problems downstream. The bottom line of this was we didn't see power gains that the increased airflow indicated we should, which was the result of mixture quality problems we'd created by just going for more air. The old story about 'more air isn't always more power' is true in a lot of cases where these gains upset combustion efficiency.

"I guess if I had to summarize all this is, I would say you can use an airflow bench for more than just measuring total flow," said Wells. "At first, it may be difficult to understand how to read the patterns the dye method produces. But by comparing them to existing combustion patterns you can see on piston crowns and combustion chambers, it's possible to figure out what needs to be done to improve combustion efficiency and power. Especially if you use dimples in areas where fuel is separating and collecting on these surfaces, the dye will show you the areas where the dimples should go. Even if you don't use dimples, these patterns will give a picture of what's changing when modifications are made to intake manifold runners, plenum chambers, intake ports, and combustion chambers. In fact, just about anything that affects the quality of air delivered to the combustion space can be included by this method."

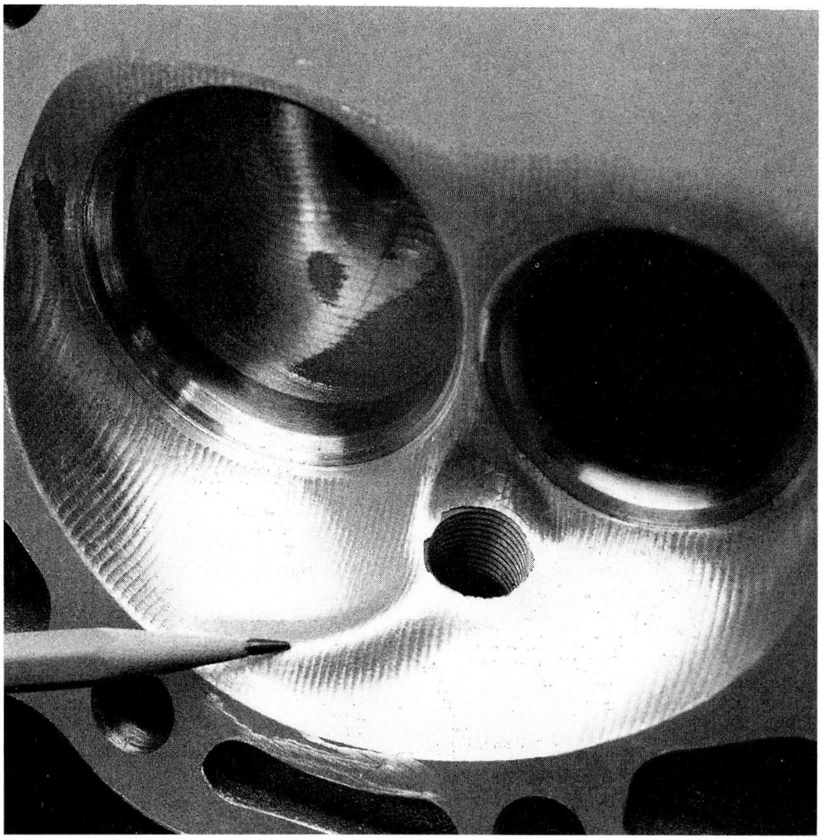

Chapter 23
ENGINE AIRFLOW

It's More Than Just Numbers

Text and Illustrations by Jim McFarland

Modifications to the back wall of the combustion chamber, such as the one shown, can help maintain the proper flow of air needed for optimizing air/fuel mixtures.

A prominent engine builder was asked, "What's the most frequent question you get when discussing cylinder heads with a customer?" His answer? "How much air will they flow, or how many cubic centimeters are the ports?"

Stop right here and take note. Nothing was said about port shape, surface finish, airflow quality, cross section versus torque range, combustion chamber shape, or any other characteristic of potential impact on optimizing power—only the amount of airflow and the size of the ports. What's implied is that more flow must mean more power and/or the bigger the better when, in fact, both expectations can be misleading, if not incorrect.

ROLE OF THE FLOW BENCH

This is not a magical piece of equipment; its limitations need to be recognized. In particular, simply increasing airflow volume (based on flow bench readings) does not always result in power gains—it can, but not conclusively. This will be discussed more in length later.

What it can do is provide a benchmark for determining the "potential" for increased horsepower. Rather than theorize on how this can be done, here's a quote from Harold Bettes, a reliable and knowledgeable technical source at SuperFlow . . . and longtime gearhead:

First of all, flow benches can be used to quantify information you've heard or been told about. Knowing airflow values is useful in order to make proper selection of parts that support airflow, such as camshafts and intake manifolds. In fact, even rear-gear ratio choices can be tied to intake airflow because peak power tends to occur at peak airflow, and this can be computed.

Here's how that works: Let's assume you've obtained intake port (or port/runner) flow data at 10 inches of water. Based on optimum spark timing, air/fuel ratio, volumetric efficiency, and burn efficiency, we can estimate a conservative level of power. Simply multiply the flow reading by 0.43 to obtain a value for power per cylinder. For example, based on a flow reading of 200 cubic feet per minute and an eight-cylinder engine:

200 cfm x 0.43 = 86 x 8 cylinders = 688 hp @ peak power rpm

If the flow data had been collected at 25 inches of water, the multiplier would be 0.27 instead of 0.43, resulting in 432 hp at peak power rpm. To determine the rpm at which these estimated values of peak horsepower would occur, a similar approach can be taken. That equation is as follows:

Peak power rpm = 2,000/displacement of one cylinder x [intake flow in cfm at 10 inches of water bench depression]

If we assume some typical values for cylinder volume and intake flow, the equation is:

Peak power rpm = 2,000/43.75 x 156 cfm = 7,131 rpm

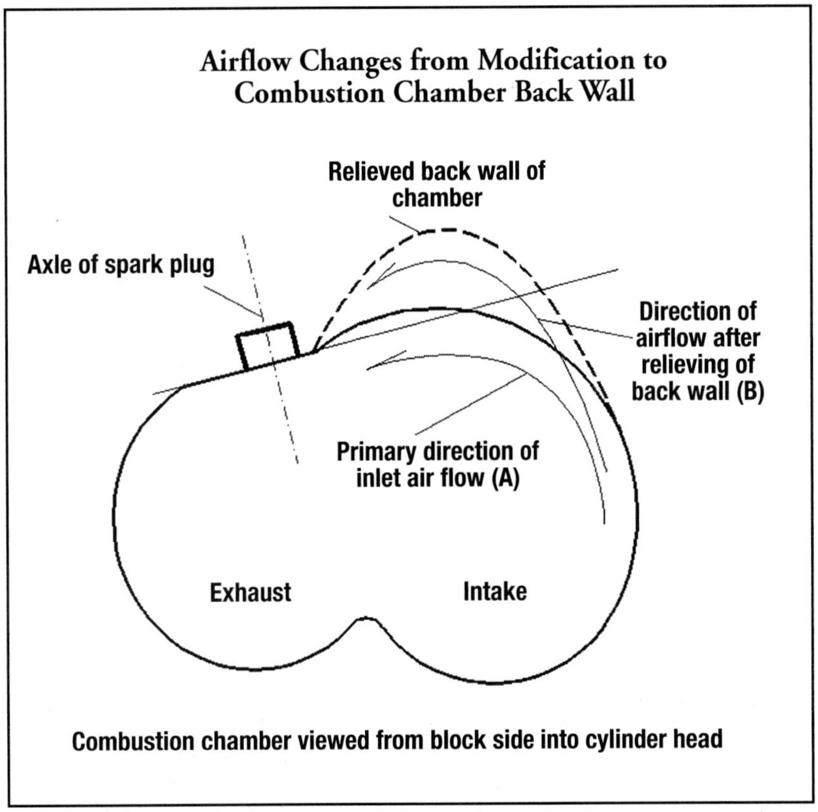

While modifications of this type may lead to net increases in inlet airflow, subsequent disruption of mixture quality can have a negative effect on combustion efficiency, the result of separated air and fuel. Evidence of this problem can be found by "spray pattern" tests using an airflow bench and valve set at about 70 percent of net lift.

Bettes emphasizes that this approach (as outlined in SuperFlow's instructional materials) assumes airflow data is obtained from a "full intake system" (carburetor, manifold, and cylinder head) and is for engines that are well sealed and of a mechanical compression ratio greater than 9:1. You should note that none of the foregoing information makes a single mention of port volume, only flow quantity.

THE IMPORTANCE OF AIRFLOW QUALITY

Aside from pure, steady-state, one-directional airflow numbers, air benches can be used to identify airflow "quality" conditions. A distinction should be made between flow quantity and quality, especially for carbureted engines. These rely upon the transfer of air and fuel from carburetor to combustion space, often plagued with problems that reduce how well the mixture is maintained its homogeneity. By proper study on a flow bench, the cause of these problems can be identified and often removed. Once again, here's how that works.

Consider air to behave like a compressible fluid. It has viscosity, effectively increasing with flow rate (speed). But it also tends to adhere to the inner surfaces of the inlet path (see illustration). Although there are exceptions to creating and maintaining stable airflow with minimum turbulence, it is usually sensible to take this approach for the majority of engine applications. But even if there are deliberate steps taken to create specific air or mixture motion (either controlled or random), air remains the "working fluid" by which fuel is communicated to the combustion space. It is therefore worthwhile to understand the effects and benefits of controlling or affecting airflow quality.

How can a flow bench be used to study airflow quality? Well, some veteran bench operators rely on the "sound" of port or passage performance. If it's "quiet," then it's stable or exhibits minimum turbulence. By the use of pressure probes, it's possible to do a more accurate job of locating and quantifying trouble spots along a flow path.

Two such examples of probes are provided in accompanying illustrations. While "velocity" probes are suitable for mapping pressure distribution patterns and identifying areas of high and low dynamic pressure in a passage, "J" probes help determine where boundary layer disturbances lie and the potential for fuel to separate from the air stream. Both of these methods are useful tools when evaluating flow passages for airflow quality. They are also of benefit when studying airflow characteristics during flow directional changes, particularly in inlet paths where fuel and air tend to separate due to the compressibility and mass (weight by volume) differences between fuel and air.

To illustrate how this weight difference can affect mixture homogeneity, consider this example: A train comprised of loaded cattle cars is moving down its tracks at a constant speed. In this example, the train is analogous to air and the cattle to fuel. At this point, the combination is well "mixed." Now, visualize what happens if the train suddenly changes in speed or direction, or accelerates or decelerates. It doesn't require much imagination to realize how the quality of this "mixture" will be affected, accordingly. Such is the case between air and fuel in an inlet path during changes in air direction or speed. The greater the difference in fuel particle size (atomization efficiency), the greater the problem of air/fuel separation, and the worse the quality of mixture delivered to the combustion space. It follows that impaired mixture quality translates into reduced power.

Surface texture is another factor that can influence flow quality, particularly on the inlet side where attempts to keep fuel suspended are made.

Roughness is especially effective in areas of low flow activity where fuel and air tend to become separated. Smooth intake ports may tend to increase raw airflow, but you can expect them to be less effective in maintaining fuel suspension than rougher surfaces.

Exhaust ports, on the other hand, respond well to polished surfaces since mixture quality is not an issue. But what is most important to remember is that flow quality cannot be overlooked in deference to quantity.

Before we leave the subject of mixture quality, it might be worthwhile to mention that in some cases it's of value to create vigorous mixture or airflow motion if the results lead to increased fuel atomization. For example, let's consider the limitations placed on "restricted" engines: those for which limited airflow and/or mechanical compression ratio is required. In these instances, it's important to burn every possible "drop" of induced fuel. So if an inlet path can be created (given the fact total flow is limited) that creates very aggressive "agitation" of the air, the possibility exists for reducing net fuel droplet size at the time of combustion, trading off flow loss to improved combustion efficiency. The more fuel that's burned, the higher the net cylinder pressure and output, giving an engine the impression its compression ratio is higher than it actually is. Little wonder some current restrictor engines are using "spiral" inlet passages.

THE INFLUENCE OF COMBUSTION CHAMBER SHAPE

It is one thing to work toward port configurations that produce sufficient flow capability, but how flow is directed into the combustion space is quite another. Taken at face value, combustion chambers are intended to work in conjunction with flow patterns created by intake and exhaust. Particularly on the intake side, dynamic patterns of air movement exiting intake passages and entering the combustion space must coincide with the geometry of the chambers. Should each of these two areas not match the functions of the other, disruption in mixture quality and combustion efficiency often result.

Stated another way, combustion chambers should be designed (or modified) to enhance or continue providing the flow benefits set up by intake (or exhaust) ports. By way of example, take a moment to examine the illustration on page 140 that shows a generic combustion chamber. Note that this chamber is laid out to help direct incoming air (or air/fuel mixture) in a counterclockwise fashion,

Flow Path Differences Between Air & Fuel in a Turn

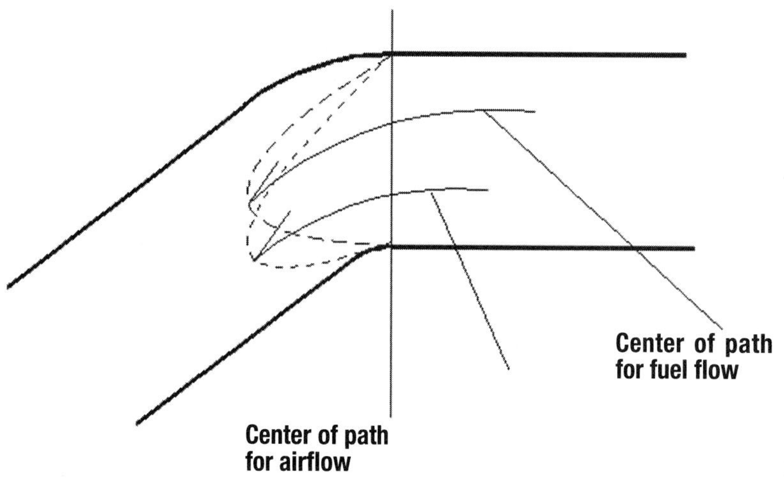

Unless accompanied by proper design of passage cross section, air and fuel tend to separate during flow direction changes. By the combined use of "velocity" and "boundary layer" probes, passages can be mapped to determine the severity of this condition and when it has been corrected. In particular, along the short side of paths changing direction, the boundary layer probe will identify areas of layer separation and potential locations where separated fuel tends to migrate. The use of so-called "D" or "trapezoid" cross sections can help bring the two centers of flow {air and fuel} more into coincidence.

Two Methods for Evaluating Port Flow Quality
(Methods can be used for areas other than in intake passages)

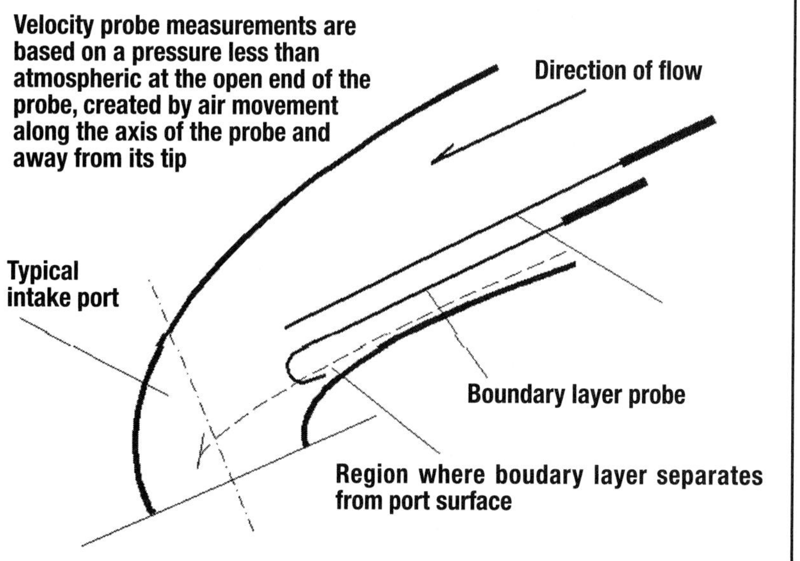

Manometric measurements of boundary layer "separation" from port surface are indicated in a manometer reading opposite to that of velocity pressure measurements. Boundary layer disturbances can lead to flow quality problems.

Static Pressure Measurement for Distribution Pressure Profiling

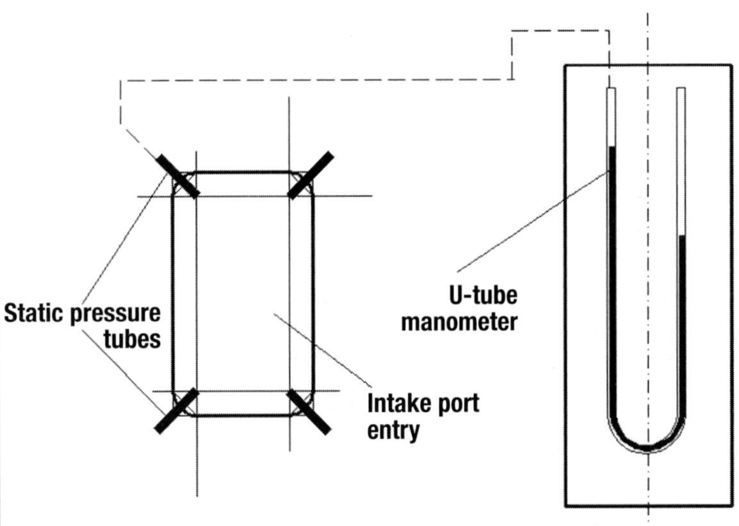

See sidebar on Pressure Profiling (right) for details on this illustration. Although only one U-tube manometer is shown, a group of four such instruments can be used for simultaneous measurement of static pressure in each of an intake port's entry corners.

thereby helping maintain the energy while aiding motion in a circular or "swirling" motion. If, for reasons of improving raw airflow (as might be measured on a flow bench), the back wall of the chamber is extended beyond the original contour, it's possible to increase total flow and decrease combustion efficiency.

According to Pete Incaudo of CNC Cylinder Heads, "There's a critical relationship between back wall position and top cut on the intake valve seat. The idea is to maintain airflow directed around toward the center of the bore while laying back the wall to improve flow. You don't want to lose the 'rotation' of incoming air, with respect to the cylinder bore. Directing too much airflow toward the back wall can destroy proper mixture motion. In an extreme case, the wall can be laid back so far that any influence by the top cut to help maintain circular motion of the inlet flow is lost. When this happens, any gains in flow from laying back the wall is lost to poor combustion efficiency. Power suffers accordingly."

Here's another instance where airflow quality becomes more important than quantity. It also reflects the fact that a mere increase in flow may not provide an attending boost in power. Unless additional airflow is constructed in a fashion that

PRESSURE PROFILING

Intake manifold runners are extensions of cylinder head inlet ports. This is one axiom among many airflow parts designers. In order to accomplish this objective, there are several methods. Among them is the matching of intake port pressure-distribution patterns to the exit flow distribution patterns of intake runners. Here's an example of how this works:

Assume there is an inlet port fitted with an entry radius, typically about 0.500 inch. Using a 0.125-inch-thick piece of plastic with an opening that matches the intake port, fitted with static pressure tubes at each corner of the opening (see illustration), a "distribution" of pressure (corner to corner) can be measured. A simple U-tube manometer can be used in lieu of something more sophisticated. Incidentally, such measurements should be made with the entry radius in place.

Remove the entry radius and bolt up the intake manifold to be used. Recheck the static pressure measurements at each corner of the inlet port. If they remain within 3 to 5 percent of their readings prior to installation of the manifold, an acceptable "pressure profile" has been determined. In some instances, a smaller percentage than this is desirable.

However, if the readings fall outside this limit, airflow quality—and often quantity—has been disrupted by installation of the manifold. In most instances, adjustment of intake manifold runner entries (either length or radius) can be modified to bring the profile into the desired range. The objective here is to prevent any "conflict" in airflow transition from intake manifold runner exit to intake port entry. As a rule, the greater the disruption (in inequality) of the pressure profile, the greater the loss in airflow quantity and quality. In particular, in carbureted engines involving wet intake flow (air and fuel), such disruptions can lead to mechanical separation of fuel from the air and unwanted changes in air/fuel ratio in the combustion space. You know the rest.

supports combustion, it becomes wasted. As intake air enters the combustion space, there is a combined "venturi" effect created between the intake valve head, valve seat, and sidewall of the combustion chamber. Consequently, the dimensional relationship between these three components is important to both net inlet flow and flow direction around the perimeter of the cylinder.

As further evidence of the importance of chamber back wall position and shape, note the photo showing a "rib" or flow "director" placed and configured to help direct air (air/fuel charge) back toward the spark plug. This same rib, another Incaudo creation, also tends to aid airflow tumble and mixture homogeneity just past its location. Brief atomized shots of machinist blue into a flowing intake port can produce chamber patterns leading to the positioning and shaping of such ribs, or other devices intended to aid flow quality in the combustion space.

COMMON FLOW BENCH OPERATOR MISTAKES

Whether you perform your own airflow tests or have the work done by a service, there are some topics and precautions worth keeping in mind. Because cylinder head and intake manifold modifications are part of his everyday activities, Dennis Wells (Wells Racing Engines) shared some of his experiences:

"If I had to list the top 10 suggestions I'd make to someone using a flow bench, here's what they'd be:

• Always check the bench's calibration before performing a test.
• Begin every flow test by checking the test piece for external leaks. On cylinder heads, this includes forgetting to install a spark plug. Just about any shop with a flow bench has been guilty of this at one time or another.
• When flowing an intake port, incorrect readings will result if the exhaust valve is pulled slightly open as a result of a weak flow spring.
• To avoid incorrect readings, make certain the flow-range control is properly set.
• Intake port entry radius must match the mouth of the port exactly, or turbulence will invalidate readings.
• The fixture attaching a cylinder head to the flow bench must be the same size as the engine's bore.
• Do not continue to enlarge ports just because flow continues to increase. You can end up with a high-flowing, large-runner-volume head that works well on the flow bench but has poor throttle response on the track. Remember, you're not racing flow benches.
• When a customer asks, 'How much cfm does it flow?' I tell them the valve doesn't remain very long at maximum lift. Be more interested in flow numbers from 0.200 to 0.600 inches of lift.
• Remember that the shape of the port and combustion chamber will usually have more impact on performance than large flowing heads.
• Spend some time learning to use velocity and boundary layer probes to study airflow quality and airflow stability. Both these topics can lead to increased power and on-track acceleration."

On the subject of where in the lift range port flow is most critical, consider the following thoughts. For both intake and exhaust valve movement, we are dealing with a continually moving valve (other than at peak lift) once it gets into its lift cycle. The idea is to determine the percentage of total intake or exhaust valve lift at which these two valves remain off their respective seats most of the time. We also need to relate these two percentages (one for the intake, one for the exhaust) to piston speed (acceleration) or crankshaft angle.

From another perspective, we're trying to figure out the valve lift at which the piston is doing the most work toward creating a low-pressure condition in the cylinder (intake stroke) and the lift at which the piston is doing the most toward ridding the cylinder of exhaust gas (exhaust stroke). Over time, it appears that these two conditions occur at about 70 percent of net intake lift and 85 percent of exhaust valve lift. In other words, other than viable concern for low lift flow (in both directions) for intakes and low lift flow (in both directions), and for exhausts, these two percentages are related to intake and exhaust valve lifts for which the ports should be the most efficient. Spend a little time with these statements. They can be worth some power.

Chapter 24

MAKE EVERY CYLINDER COUNT

Engines Make the Most Power When Every Cylinder Is Individually Tuned

Text and Illustrations by Jim McFarland

Let's get right to the heart of this matter. Visualize a single-cylinder, four-stroke engine. It has an intake manifold, a carburetor, an exhaust pipe, a camshaft, and an intake and exhaust valve. Simple enough, right? We know it's possible to manipulate this little engine's torque curve over a range of rpm by adjusting the flow path diameter and length in combination with valve timing. That's a given. But first, we need to discuss some background information before diving into procedures that optimize individual cylinder power.

THE BACKGROUND

We know that the net torque curve produced by our engine is a combination of an induction system curve and an exhaust system curve. We can tune these two curves separately and combine them to locate overall torque precisely where we want it to be.

While flow activity in an intake or exhaust system is unsteady (neither constant nor linear), a mean flow velocity (m.f.v.) can be determined at any particular rpm. Factors that influence m.f.v. are piston displacement, compression ratio, valve timing, and rod-to-stroke geometry. At peak torque, there is an m.f.v. that's basically identical for all engines. It lies in the range of 240–260 feet/second.

Of the factors that impact m.f.v., the cross-section area of the flow path is critical. All else being equal, change this area, and the m.f.v. changes correspondingly. At any given rpm, an increase in the cross-section area will result in a decrease in flow rate. Decrease this area, and the flow rate increases.

Based on these relationships, the design or selection of intake and exhaust components can materially affect where an engine produces its torque, relative to rpm. We know that an m.f.v. of 240–260 feet/second occurs at peak torque, so we can use this information as a tool in the sizing (or choice) of intake manifolds and headers. We'll get more into that a bit later.

So far, we've discussed the characteristics of a single-cylinder engine. Now let's see what happens if we add cylinders. Two factors immediately become apparent. One is the influence on piston movement and torque transfer when two or more cylinders are connected to the same crankshaft. One cylinder's power strokes can be felt among cylinders that are not on their own power strokes. Crankshafts tend to wrap up during this process and experience oscillations throughout their length. This disturbs piston strokes in adjoining cylinders and affects their volumetric efficiency.

Simply stated, a spark-timing figure that may be correct for a cylinder at the front of an engine may need to be retarded for cylinders toward the rear. This torsional deflection of the crank requires individual cylinder spark timing if power is to be optimized. Valve-timing accuracy also is upset, according to the amount of crankshaft (and camshaft) wrapup.

Another significant factor is the so-called "cross-talk" (pressure excursions and disturbances) that occurs in the intake and exhaust systems of joined cylinders. Visualize a single-plane intake manifold where all port runners connect to a common plenum chamber. Pressure pulses traverse the chamber throughout the engine's rpm and load range and tend to influence pure inlet flow among the cylinders. Depending on the amount of cylinder pressure at each intake valve opening, these disturbances can reduce each cylinder's ability to optimize power.

They also can contaminate fresh air/fuel mixtures and disrupt each cylinder's ability to achieve its highest level of volumetric efficiency. Inspection of piston tops and combustion chambers can reveal specific burn patterns as well as the power produced by individual cylinders. More uniformly distributed burn patterns are characterized by a lack of light and dark combustion residue on pistons and chambers. Concentrations of dark patterns suggest overly rich burns and lost power, especially at higher rpm.

Clearly then, net power is produced by the combination of an engine's components, so it is important to recognize the relationships among them. Understanding their integrated working is vital to selecting and matching parts intelligently. Let's look at some of these pieces.

INTAKE MANIFOLDS

Manifold runner lengths are typically unequal, particularly for carbureted engines. As with unequal header primary pipe lengths, this inequality affects the relationship between torque output and rpm. Since Edelbrock introduced its single-plane, V-8–type manifold in the early '70s, tens of thousands of similar design concepts have proliferated in the circle-track community. Yet the camshafts most often used with these manifolds do not compensate the short runners for the inner four cylinders and long runners for the outer four. Some Nextel Cup teams have discovered the benefit of such camshafts; the Saturday-night contingent has yet to catch on.

EXHAUST HEADERS

The use of equal (or unequal) length primary pipes and the incorporation of one or more stepped

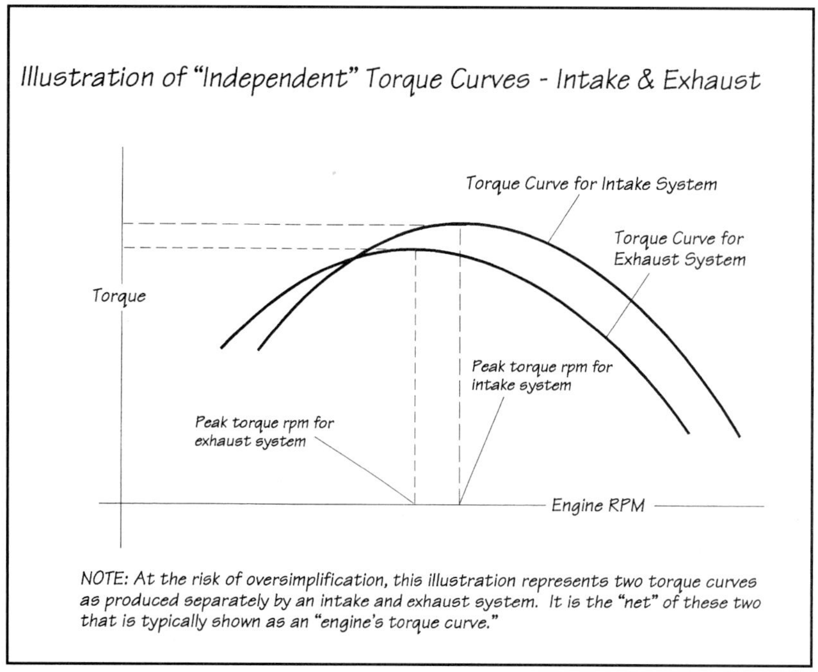

Illustration 1: Intake and exhaust systems make separate contributions to overall torque output. Therefore, each is a candidate for independent tuning. The wider apart each is tuned, the flatter the net torque curve. The closer each is tuned, the more the aggregate torque curve peaks.

This single-plane Edelbrock manifold is typical of designs incorporating runners of different length. It offers an opportunity to tune other components of the engine to vary the point at which peak individual cylinder torque is produced.

sizes of pipe are common among circle-track engine builders. What may not be fully understood is the control each of these features provides in the placement of torque relative to rpm.

The concept of equal-length header pipes is a proven benefit. However, the assumption is that each connected cylinder is making the same power as any other, which is usually incorrect. Variations in combustion efficiency between cylinders, flame travel, and exhaust contamination tend to prevent

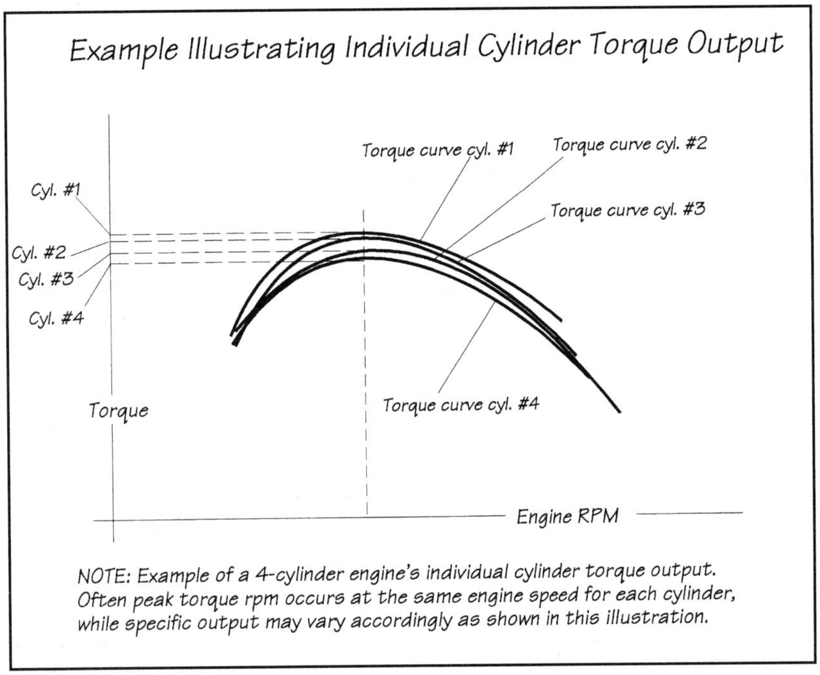

Illustration 2: This series of plots shows how torque output and peak torque rpm points can vary among an engine's cylinders. The peaks may not occur at the same rpm, which creates a further broadening of the net torque curve.

individual cylinders from making identical power. Note this later in the section discussing engine cycle analysis (ECA).

CYLINDER HEADS

In most cases, individual combustion chambers do not produce identical burn patterns or pressure. Variation in chamber shape is one cause, but the most frequent culprit is an inequality of air delivered by individual manifold runners.

Remember, the total inlet path is comprised of both the intake manifold and the intake port. So it is sometimes necessary to adjust the port shapes to optimize combustion pressures in each combustion space. Burn patterns in the chambers or on the piston tops will enable you to achieve this, or it can be done by measuring in-cylinder combustion dynamics via engine cycle analysis or similar methods. The limited budget of most Saturday-night racers confines them to visual inspection of components.

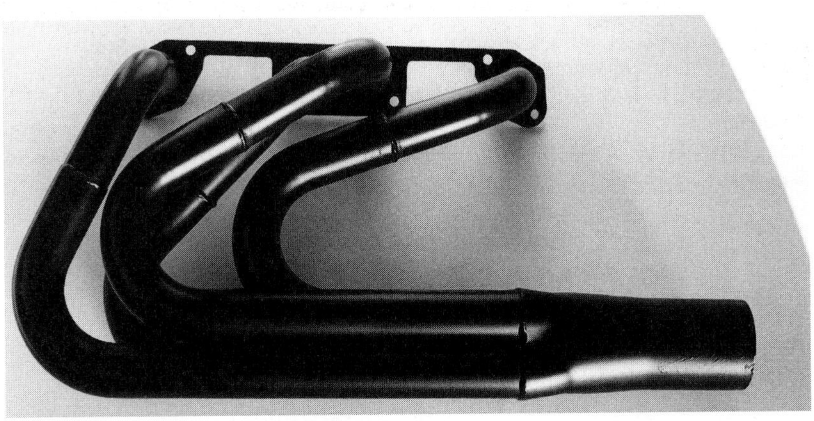

Hedman Hedders' stepped primary header pipe is an example of how each cylinder's torque output can be broadened over a wider range of rpm.

CAMSHAFTS

Manufacturers are increasingly happy to provide custom grinds, so consider having intake and exhaust lobes cut to match the rpm ranges of different inlet paths. For example, the two-length intake manifold design previously described can produce a broader torque band by incorporating an intake and exhaust lobe for the long (or bottom-end torque) length and another intake and exhaust lobe for the short (or higher rpm) length. It doesn't require much imagination to see how this works; in fact, the practice could be incorporated into header-pipe sizing that matches the intake paths to which each pipe is attached (illustration 3).

IGNITION SYSTEMS

Several manufacturers, including MSD, offer systems that permit individual cylinder-spark timing. Settings can be determined by combustion-process feedback that keys off ignition spark ionization at the plug's gap. Cylinder-to-cylinder air/fuel ratios vary as a function of rpm, load, and mixture-motion characteristics, so the ability to optimize spark timing for each cylinder is an additional tool for maximizing total power. We've even seen electronic ignition systems with trigger wheels (reluctors) of uneven tooth spacing that vary the amount of spark advance from cylinder to cylinder. All these techniques address the specific spark-timing requirements of cylinders displaying different pre-combustion patterns.

ENGINE CYCLE ANALYSIS (ECA)

ECA is a form of high-speed information gathering, which is useful largely because of the speed at which the internal-combustion process takes place. The process measures the relationship between cylinder pressure and crankshaft angle. A stream of information is gathered, showing a pressure history of what occurs inside the cylinder. From this basic information, a series of mathematical computations are made to determine other combustion-related performance characteristics. That's the 25-cent definition.

Specifically, an encoder is fitted to an engine's crankshaft snout and is electronically linked to one or more pressure-sensing devices exposed to the combustion pressure in a single cylinder. That was a decade ago. Today, all of an engine's cylinders may be analyzed simultaneously and are sometimes coupled with pressure sensors in the intake and

Backpressure reductions also allow individual cylinder tuning to be optimized. This Flowmaster "Delta Flow" concept is reported to provide low-pressure conditions for additional cylinder evacuation and increased power.

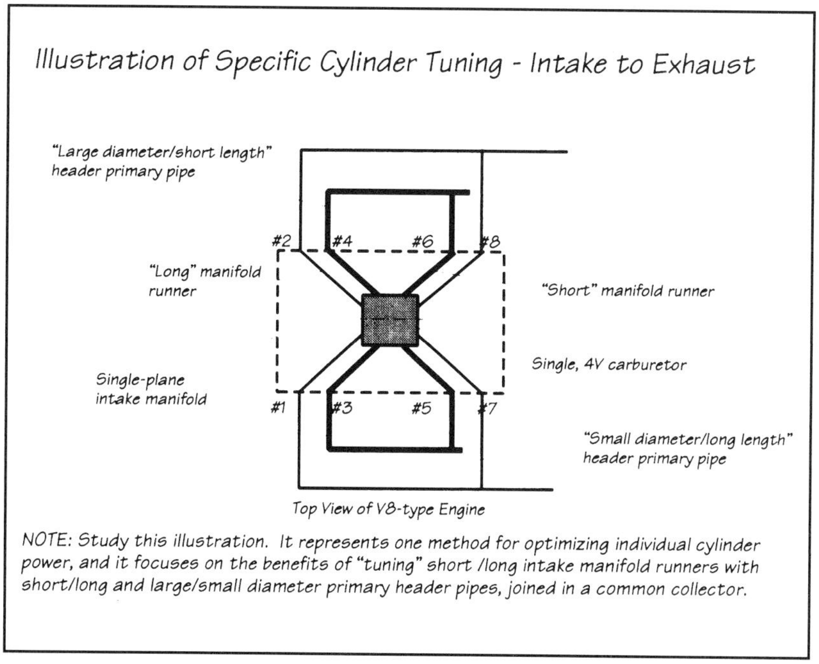

Illustration 3: Matching intake and exhaust path lengths (including cylinder-specific valve tuning) is particularly effective when attempting to optimize short- and long-track torque requirements.

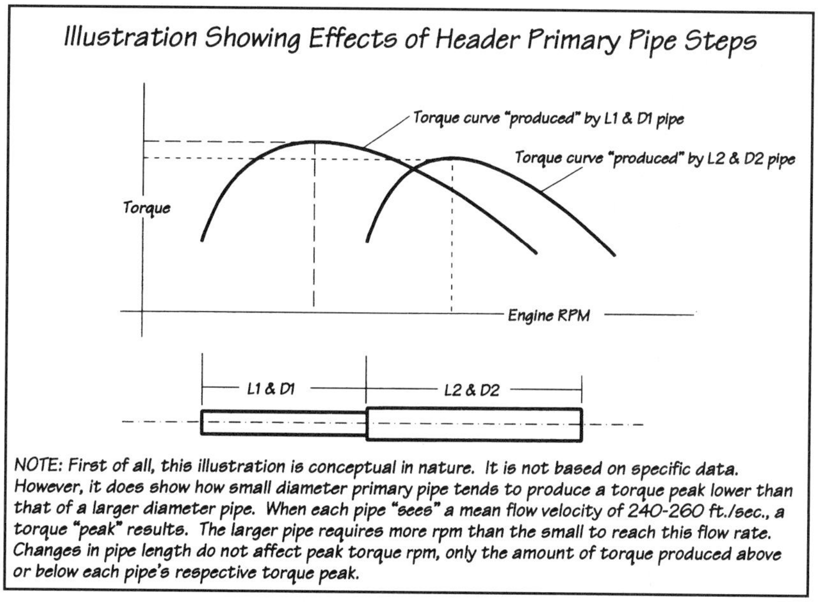

Illustration 4: Stepped headers produce a broader and flatter torque curve. Make certain to match the rpm and torque requirements with the proper primary pipe size and step location.

exhaust tracks of each cylinder.

ECA studies power levels produced by each cylinder over a range of engine speeds and loads. Data acquisition can occur at a rate of one million samples per second and crank angles down to one tenth of a degree, producing almost real-time information. Components such as manifolds, ports, combustion spaces (chambers and piston tops), valve timing, and spark timing can be matched to make all cylinders produce identical power. The process is especially useful in determining the best compromises for power in circumstances when cylinders "talk" to one another through common areas such as intake manifold plenums and exhaust header collectors.

It is important for the Saturday-night racer to be aware of this technology. It's only a matter of time before it becomes affordable from engine dyno operators as a cost-effective way of evaluating customers' engines. Then it will then find its way into the mainstream of circle-track racing. It is already in use among leading professional teams, especially in Nextel Cup, IRL, and Formula One. For more information, see the box beginning on page 149.

OPTIMIZING POWER

Hi-Techniques, based in Madison, Wisconsin, is a leading source for ECA, and the company has enlightened some widely recognized teams. See the accompanying engine-dyno data that shows the effects of optimizing spark timing and other engine variables on an individual cylinder basis (especially valve timing and compression ratio).

One of the best approaches to optimizing individual cylinder power is to consider certain physical aspects of your components and tune them accordingly. For example, connecting short and long intake runners to large and smaller primary header pipes helps broaden the torque curve (see illustration 3). Selecting valve timing that is consistent with these two ranges of torque output can further enhance a broader, flatter torque curve.

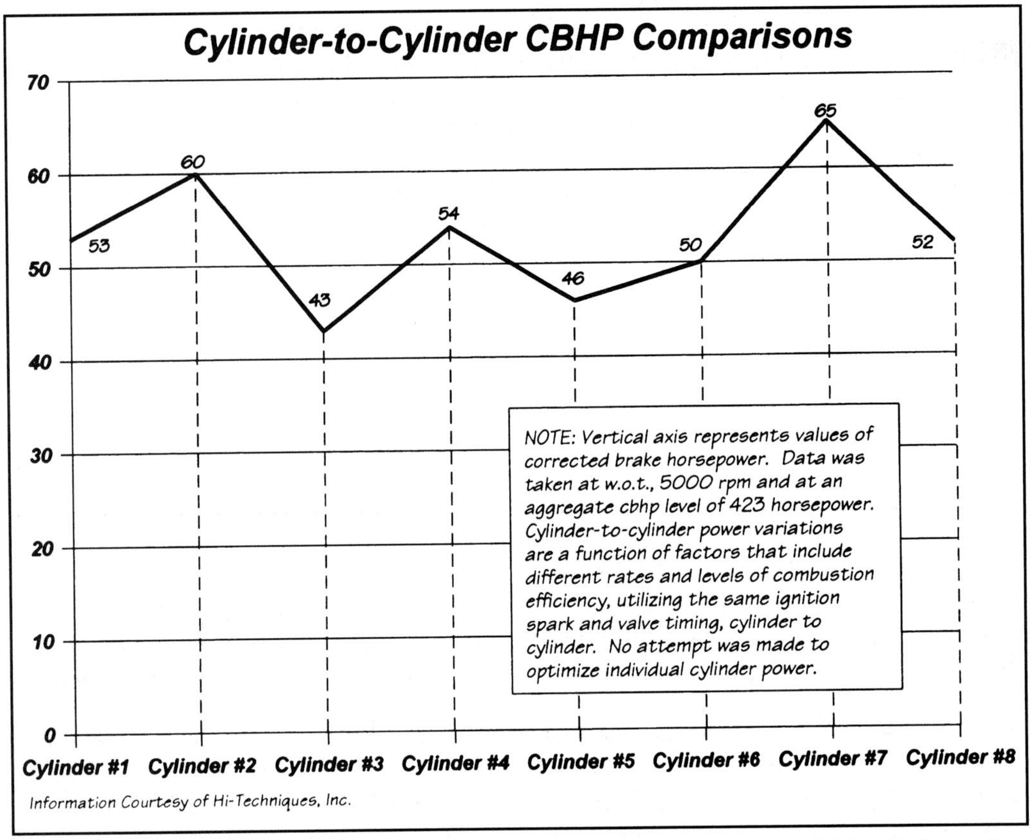

Illustration 5: These traces represent the in-cylinder pressure histories of an eight-cylinder engine. Variations in peak pressure (vertical axis) and the different crankshaft angles at which these peaks occur (horizontal axis) are further evidence of cylinder-to-cylinder power differences, as measured by engine cycle analysis.

According to Hedman Hedders President Ron Funfar, "We've found particular benefit in the 'stepped header' concept. Race headers, as well as those for light- and medium-duty trucks used for towing, show torque gains in the low- and mid-range rpm when built with properly sized primary tubing."

Using the rule that 240–260 feet/second of flow velocity in an exhaust or intake path equates to peak torque rpm, the following equation is useful in selecting these passages' sizes. There are fundamental differences between intake and exhaust flow (including gas/mixture temperatures and piston position at peak-flow velocity), but you can still arrive at workable dimensions by using this mathematical approach:

Peak torque rpm = (Passage cross-section area x 88,200) ÷ cylinder volume

By transposing these terms, you can determine the rpm at which a given cross section will produce a torque peak (intake or exhaust) or what cross-section area is required to produce peak torque at a desired rpm. The equation is a tool that can be used to evaluate existing cross sections or determine cross sections required to produce specific torque peak rpm points. Plug in a few numbers—you'll quickly see the value of this equation.

So, what about spark timing? Cross-jetting carburetors to compensate for poor cylinder-to-cylinder air, fuel, or air/fuel distribution is a time-honored method of tuning for best power. But even when plug readings begin to equalize, that does not necessarily mean that each cylinder's power has been optimized; it simply means the burn appearance of the plugs is similar. Advancing or retarding the spark timing of individual cylinders is a step toward compensating for variations in the volume of air and fuel in any given cylinder. For example, advancing the timing for a cylinder showing a slightly rich condition may indicate that more timing would help burn the fuel that's already present, resulting in a power increase for that particular cylinder.

The use of stepped headers also helps to extend torque curves, aiding off-the-corner torque while still retaining straightaway speed. How this works is shown in illustration 4. Of course, the selection of pipe size and step location will affect where and how much torque is shifted between the two peaks derived from the pipe outside-diameter (o.d.) selection. Here's how that works.

We have previously shown that the peak torque rpm point for a given primary pipe is determined by pipe area (a function of o.d. and wall thickness). Changes in pipe length affect how much torque is delivered above and below this peak. Shortening the pipe adds torque above the peak and removes it from below. The reverse is true of lengthening the pipe.

Therefore, in a given primary header pipe comprised of two pipe areas, moving the step up or down the pipe's overall length will add or subtract torque above or below the peak of either size pipe. That sounds complex, so examine illustration 4. The caption and footnote should clear up any confusion about the effects of adjusting primary pipe step location.

IS IT WORTH THE TROUBLE?

That depends upon how many ways you'd like to split the hairs. Hi-Technique's example engine—a 350ci small-block Chevy—produces a corrected 826 bhp. Optimizing individual spark timing increased this number to 836.5. Further manipulation of valve timing and compression ratio raised this to 848.5 hp. For engines built to rules that allow limited modifications, "power by the cylinder" may make more sense than if the rules are more lenient. Then again, ultra-competitive teams may wish to optimize parts functions regardless of what the rules allow. In any event, maximum engine power is not obtainable unless the potential combustion efficiency and output of each cylinder is addressed, no matter how you choose to take those steps.

THE ABC'S OF ECA

It's one thing to develop spark plug–reading skills, but this is an "after-the-fact" method of determining optimum individual cylinder power output. You're looking at the results of conditions, not their causes. As engine builders, we strive to make everything equal among cylinders dimensionally, regarding parts balance, valve timing, intake and exhaust port flow and a host of related variables.

However, once an engine is fired, a multitude of pressure pulses and excursions begin to take place, tending to upset all the painstaking work done to equalize the machined, timed and balanced parts among cylinders. In short, while so-called "blueprinting" efforts are beneficial and necessary, as is port flow equality, pulsations in intake and exhaust systems tend to upset the ability of one cylinder to produce the same power as another.

As a case in point, in support of the fact cylinders can "talk" to each other (with respect to pressure pulsations), visualize what it takes to transmit the head of a broken intake valve from a cylinder on one side of an engine to another on the opposite side ... passing through the plenum of a single-plane manifold. It happens frequently. Plenum dividers insufficiently anchored in an intake manifold can be fractured in a similar pressure-pulsing fashion. Keep in mind that these "cross-talk" pressure variations among an engine's cylinders are often comprised of elements that decrease combustion efficiency, such as exhaust gas.

There's also the issue of cylinder-to-cylinder power variations caused by air/fuel mixture quality problems. Mixture ratios can vary, based not only on contamination by exhaust gas but from improper mixing of air and fuel between where it enters the air stream and the combustion space. And while it's true that reading spark plugs provides an indication of a problem, it doesn't quantify power loss or do it by increments. Engine cycle analysis (ECA) does.

By the use of a crankshaft snout-mounted device (an encoder), a surface-mounted pressure transducer in the combustion space (or several in other pressure-critical areas such as the intake and exhaust passages . . . singular or multiple locations) and the

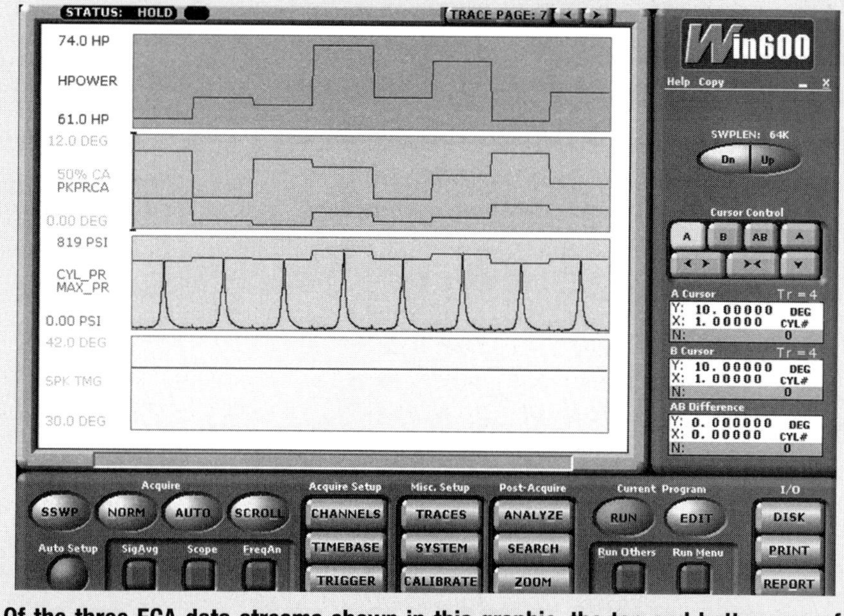

Of the three ECA data streams shown in this graphic, the top and bottom are of particular interest. Each of these reflects individual cylinder pressure histories (bottom trace) and corresponding horsepower output (top trace). Reading from left to right (cylinder No. 1 through No. 8), the horsepower numbers are as follows: 62.3, 65.4, 64.3, 72.7, 65.4, 70.6, 62.3 and 66.4. Aggregate horsepower is 529.4 as observed at 4,167 rpm. This example is intended to show how ECA can reveal not only variations in cylinder-to-cylinder output but also quantifies these differences as reference for modifications intended to equalize power among all cylinders.

proper computer software package, cylinder pressure vs. crankshaft angle data can be gathered. Such information acquisition can take place quite rapidly (500,000 pieces/second or faster) and at small increments of crankshaft angle (1/10th degree, for example). While this is a highly simplified description of the process and mechanism, it is intended to provide a fundamental understanding of the method used.

The result of this process is virtually a "real-time" data stream of information, the raw form of which presents a pressure/crank angle trace that can be related to piston position. Plus, from these data, certain thermodynamic calculations can be performed to produce a variety of combustion-related information. Typically, these calculations are included in an ECA data acquisition package. For example, you may want to contact Hi-Techniques in Madison, WI, or visit its website at www.hi-techniques.com for additional information about system specifics.

From a practical (or racer's) point of view, let's take a look at some of the ways ECA is currently being used by leading Nextel Cup racing teams in particular. Interestingly, multi-cylinder engines don't enjoy equal power output among their cylinders, as previously pointed out. At least they typically don't unless this issue is specifically addressed. ECA helps get this accomplished.

The accompanying chart depicts comparisons of individual cylinder output, measured at 4,167 rpm and a total horsepower of 529.4. From a minimum power output of 62.3 hp (cylinder No. 1) to a maximum horsepower of 72.7 (cylinder No. 4), you can note the output variations among the remaining six cylinders ... at the same rpm. While ECA doesn't tell you how to correct such problems, it can identify which cylinders are out of balance and quantify the differences.

As you might expect, there are other uses of the technology. For example, by positioning pressure transducers in the intake path, combustion space and exhaust passage, it's possible to simultaneously study (and manipulate) net cylinder pressure or power by comparing pressure "histories" before, during and after combustion . . . all as a function of crankshaft angle (piston position). Consider this benefit in the course of designing (or selecting) a camshaft that optimizes the remainder of an engine package . . . or, for that matter, configuring an entire engine package of components from scratch.

Cost-wise, ECA packages may not appear an economical venture. But when you consider the amount of money the technology can save by the reduction or elimination of expensive parts selection mistakes (and lost power), its use comes into sharper focus. And even if you cannot afford the luxury of ECA and its benefits, an increasing number of professional engine builders are incorporating the technique into customer engine configurations.

Overall, ECA points to the value of being able to measure individual-cylinder output. After almost a decade of exposure to and use of the technology, this writer can attest to the impact it can have on optimizing engine power. In fact, it's as clear as ECA.

NOTES

NOTES

GENERAL MOTORS
Big-Block Chevy Engine Buildups: 1-55788-484-6/HP1484
Big-Block Chevy Performance: 1-55788-216-9/HP1216
Camaro Performance: 1-55788-057-3/HP1057
Camaro Restoration Handbook ('67–'81): 0-89586-375-8/HP1375
Chevelle/El Camino Handbook: 1-55788-428-5/HP1428
Chevy S-10/GMC S-15 Handbook: 1-55788-353-X/HP1353
Chevy Trucks: 1-55788-340-8/HP1340
How to Hot Rod Big-Block Chevys: 0-912656-04-2/HP104
How to Hot Rod Small-Block Chevys: 0-912656-06-9/HP106
How to Rebuild Small-Block Chevy LT-1/LT-4: 1-55788-393-9/HP1393
John Lingenfelter: Modify Small-Block Chevy: 1-55788-238-X/HP1238
LS1/LS6 Small-Block Chevy Performance: 1-55788-407-2/HP1407
Powerglide Transmission Handbook:1-55788-355-6/HP1355
Rebuild Big-Block Chevy Engines: 0-89586-175-5/HP1175
Rebuild Gen V/Gen VI Big-Block Chevy: 1-55788-357-2/HP1357
Rebuild Small-Block Chevy Engines: 1-55788-029-8/HP1029
Small-Block Chevy Engine Buildups: 1-55788-400-5/HP1400
Small-Block Chevy Performance: 1-55788-253-3/HP1253
Turbo Hydramatic 350 Handbook: 0-89586-051-1/HP1051

FORD
Ford Windsor Small-Block Performance: 1-55788-323-8/HP1323
Mustang Performance (Engines): 1-55788-193-6/HP1193
Mustang Performance 2 (Chassis): 1-55788-202-9/HP1202
Mustang Performance Engine Tuning: 1-55788-387-4/HP1387
Mustang Restoration Handbook ('64–'70): 0-89586-402-9/HP1402
Rebuild Big-Block Ford Engines: 0-89586-070-8/HP1070
Rebuild Ford V-8 Engines: 0-89586-036-8/HP1036
Rebuild Small-Block Ford Engines: 0-912656-89-1/HP189

MOPAR
Big-Block Mopar Performance: 1-55788-302-5/HP1302
How to Hot Rod Small-Block Mopar Engine Revised: 1-55788-405-6/HP1405
How to Maintain & Repair Your Jeep: 1-55788-371-8/HP1371
How to Modify Your Jeep Chassis/Suspension for Offroad: 1-55788-424/HP1424
How to Modify Your Mopar Magnum V8: 1-55788-473-0/HP1473
How to Rebuild Your Mopar Magnum V8: 1-55788-431-5/HP1431
Rebuild Big-Block Mopar Engines: 1-55788-190-1/HP1190
Rebuild Small-Block Mopar Engines: 0-89586-128-3/HP1128
Torqueflite A-727 Transmission Handbook: 1-55788-399-8/HP1399

IMPORTS
Baja Bugs & Buggies: 0-89586-186-0/HP1186
Honda/Acura Engine Performance: 1-55788-384-X/HP1384
Honda/Acura Performance: 1-55788-384-X/HP1384
How to Hot Rod VW Engines: 0-912656-03-4/HP103
Mitsubishi & Diamond Star Performance Tuning: 978-1-55788-496-1/HP496
Porsche 911 Performance: 1-55788-489-7/HP489
Rebuild Air-Cooled VW Engines: 0-89586-225-5/HP1225
The VW Beetle: A History of The World's Most Popular Car: 1-55788-421-8/HP1421
The Volkswagen Super Beetle Handbook: 1-55788-483-8/HP1483

HANDBOOKS
Automotive Detailing: 1-55788-288-6/HP1288
Auto Electrical Handbook: 0-89586-238-7/HP1238
Auto Math Handbook: 1-55788-020-4/HP1020
Automotive Paint Handbook: 1-55788-291-6/HP1291
Auto Upholstery & Interiors: 1-55788-265-7/HP1265
Car Builder's Handbook: 1-55788-278-9/HP1278
Classic Car Restorer's Handbook: 1-55788-194-4/HP1194
Engine Builder's Handbook: 1-55788-245-2/HP1245
Engine Cooling Systems: 978-1-55788-425-1/HP1425
Fiberglass & Other Composite Materials Rev.: 1-55788-498-6/HP1498
The Lowrider's Handbook: 1-55788-383-1/HP1383
Metal Fabricator's Handbook: 0-89586-870-9/HP1870
1001 High Performance Tech Tips: 1-55788-199-5/HP1199
1001 MORE High Performance Tech Tips: 1-55788-429-3/HP1429
Paint & Body Handbook: 1-55788-082-4/HP1082
Performance Ignition Systems: 1-55788-306-8/HP1306
Pro Paint & Body: 1-55788-394-7/HP1394
Sheet Metal Handbook: 0-89586-757-5/HP1757
Welder's Handbook Revised: 978-1-55788-513-5/HP1513

INDUCTION
Holley 4150: 0-89586-047-3/HP1047
Holley Carbs, Manifolds & F.I.: 1-55788-052-2/HP1052
Rochester Carburetors: 0-89586-301-4/HP1301
Tuning Accel/DFI 6.0 Programmable F.I.: 1-55788-413-7/HP1413
Turbochargers: 0-89586-135-6/HP1135
Street Turbocharging: 1-55788-488-9/HP1488
Weber Carburetors: 0-89586-377-4/HP1377

RACING & CHASSIS
Bracket Racing: 1-55788-266-5/HP1266
Chassis Engineering: 1-55788-055-7/HP1055
4Wheel & Off-Road's Chassis & Suspension: 1-55788-406-4/HP1406
Dirt Track Chassis & Suspension: 978-1-55788-511-1/HP1511
How to Make Your Car Handle: 0-912656-46-8/HP146
How to Get Started in Stock Car Racing: 1-55788-468-4/HP1468
How to Build a Winning Drag Race Chassis & Suspension: 978-1-55788-462-6/HP1462
Racing Engine Builder's Handbook: 1-55788-492-7/HP1492
Stock Car Racing Engine Technology, 978-1-55788-506-7/HP1506
Stock Car Setup Secrets: 1-55788-401-3/HP1401

STREET RODS
How to Build a 1934–'35 Chevy Street Rod: 978-1-55788-514-1/HP1514
How to Build a 1935–'40 Ford Street Rod: 1-55788-493-5/HP1493
Street Rodder magazine's Chassis & Suspension Handbook: 1-55788-346-7/HP1346
Street Rodder's Handbook, Rev.: 1-55788-409-9/HP1409
Street Rodding Tips & Techniques, 978-155788-515-9/HP1515